Automated Analysis of Virtual Prototypes at the Electronic System Level

Mehran Goli • Rolf Drechsler

Automated Analysis of Virtual Prototypes at the Electronic System Level

Design Understanding and Applications

 Springer

Mehran Goli
University of Bremen and DFKI GmbH
Bremen, Germany

Rolf Drechsler
University of Bremen and DFKI GmbH
Bremen, Germany

ISBN 978-3-030-44284-2   ISBN 978-3-030-44282-8   (eBook)
https://doi.org/10.1007/978-3-030-44282-8

This Springer imprint is published by the registered company Springer Nature Switzerland AG.
The registered company address is: Gewerbestrasse 11, 6330 Cham, Switzerland

To Ronja, Nik and Paya

and

To Fikriye and Hayrettin

Preface

Modeling electronic systems using the *Virtual Prototype* (VP) concept has become an industry-accepted solution in the last decade. The main reason is that VPs are much earlier available, and their simulation is orders of magnitude faster in comparison to the hardware models implemented at lower levels of abstraction. Hence, VP is used as a reference model for several applications (e.g., test, debugging, verification, and design space exploration) in the design process. To handle these applications, applying simulation-based techniques is still the predominant approach where using them requires accurate knowledge of the VP structure and behavior. However, accessing this information requires analyzing a given VP which is still a very challenging and non-trivial task. This book aims to provide a set of comprehensive and automated approaches, allowing designers to handle various tasks in the design process. The book offers the VP analysis approaches (design understanding and the infrastructures and solutions to handle other applications in the design process) from two different perspectives that have not been studied so far: debugger-based and compiler-based approaches.

Bremen, Germany

Mehran Goli
Rolf Drechsler

Acknowledgments

We would like to particularly appreciate all those who have contributed to the results included in this book. Our special thanks go to Jannis Stoppe and Daniel Große for their investments of time and brilliant ideas, which were of significant importance for this book. We also would like to acknowledge Muhammad Hassan for his contributions, which helped us to improve this book. Many thanks to Alireza Mahzoon for his helpful feedback and inspiring discussions. We would like to express our appreciation to all of the colleagues in the group of computer architecture at the University of Bremen and the Cyber-Physical Systems group at the German Research Center for Artificial Intelligence for their support.

Bremen, Germany
April 2020

Mehran Goli
Rolf Drechsler

Contents

List of Figures

List of Tables

List of Algorithms

Chapter 1
Introduction

Hardware complexity nowadays is a major issue. Modern electronic circuits and systems consist of many different functional blocks, including multiple (third-party) *Intellectual Property* (IP) cores, various on-chip interconnects and memories. According to "Moore's Law," the number of components in integrated circuits grows exponentially, and doubles almost every 2 years [79]. Currently, modern embedded systems (e.g., 28-core Intel Xeon Platinum chip) are manufactured as multi-processor systems on a single chip and reach up to 8 billion transistors [112]. The increasing design complexity and scale of *System-on-Chip* (SoC) designs along with non-functional aspects and constraints on the final system such as optimal power consumption, robustness, and reliability make the design process of such complex systems crucial.

The classical *Hardware Description Languages* (HDLs) such as Verilog [82] and VHDL [81], which are widely used to model hardware at the *Register Transfer Level* (RTL) [116], cannot be used easily to run, test, or evaluate modern embedded systems at such low-level of abstraction. The main reason for this limitation is that to perform the above tasks by using these HDLs, an accurate and complete description of the whole system is required. It means that designers first have to build the whole system and then start testing or evaluating it. As a consequence, the architecture changes or modifications, software development, and overall testing process of embedded systems can only be performed once the system is fully implemented. This makes the development process very expensive and time-consuming. Therefore, the traditional HDLs increasingly face the issue of the design gap [33], i.e., the problem that a system cannot be designed in accordance with the available manufacturing capabilities due to its complexity.

One possible solution to handle the complexity of the electronic circuits is to raise the level of abstraction towards the *Electronic System Level* (ESL) [75]. The academic and research communities have developed the ESL paradigm over the last decade to provide designers with new design methodologies, tools, and specific system design languages. This enables designers to model a system using a modular

© Springer Nature Switzerland AG 2020
M. Goli, R. Drechsler, *Automated Analysis of Virtual Prototypes at the Electronic System Level*, https://doi.org/10.1007/978-3-030-44282-8_1

approach with a set of coarse structures and underlying high-level programming language (e.g., C++) to describe the detailed behavior of these parts. By this means, a system can be prototyped quickly and used as a reference model for lower levels of abstraction. Moreover, before the actual hardware is manufactured, designers can test or evaluate the software parts of an embedded system. Thus, it acts as the standard communication platform among system designers, embedded software developers, and hardware engineers along the design process.

At the ESL, the C++-based, standardized SystemC library [56] has become the de-facto standard modeling language [100] which can be used to describe embedded systems as a *Virtual Prototype* (VP). In this context, a VP is an abstract, and executable software model which is implemented using SystemC and its *Transaction Level Modeling* (TLM) [5] framework. Along with an event-driven simulation environment, the SystemC library provides a notion of timing, which is well-suited for modeling circuits. Apart from the cycle-accurate timing model (i.e., enabling designers to have RTL-based description of designs) that SystemC provides, its TLM framework allows designers to describe a system in terms of abstract communication by using a set of rules (the base protocol) and standard interfaces. TLM focuses on the functionality of data transfers among computational components and separates the details of communication from the computation. Overall, the adoption of SystemC-based VPs (due to their much earlier availability as well as the extremely faster simulation speed in comparison to the RTL models) has led to significant improvements in the design process and reduction on time-to-market constraints [20].

Although SystemC-based VPs reduce the complexity of modeling and simulation of SoC designs at the ESL, these are not the only "system-level use cases" for which a SystemC VP is implemented and used. As illustrated in Fig. 1.1, designers of such models may want to

- test new features and validate the capabilities of SoCs, i.e., to decide where to reuse existing IPs or add new (third-party) hardware blocks,

Fig. 1.1 Design process

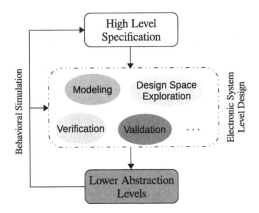

- perform design space exploration for the next generation of SoCs, and
- debug, verify, or synthesize the existing SoCs.

To handle the tasks mentioned above, accurate knowledge about VPs' structure and behavior is necessarily required. This initial step in the design process is considered as the *Design Understanding* phase. Analyzing a given SystemC VP in order to know the respective components of the design (e.g., structure), as well as their relation to each other (e.g., behavior), is a crucial and non-trivial task. Mainly because C++ (and thus also SystemC, which is a library for the former) is inherently hard to analyze due to

- the lack of proper analysis methods for C++ run-time behavior (e.g., the lack of a modern reflection framework),
- the countless compiler-specific dialects that a source code may be written in, and
- the executable binary format which is not only stripped of any information that is not needed to execute the simulation but may also be heavily optimized.

While a well-written and up-to-date documentation should be a primary source to understand the intricacies of a VP, there may be situations where these documents are poorly written, outdated or not available at all, e.g., *Third-Party IP* (3PIP) or legacy models [98]. Especially, when new designers are joined to a project or to keep other engineers working on the project up-to-date with recent changes, a lack of proper documentation exaggerates the situation. Thus, the development team members spend significant time to reverse engineering the design (e.g., reading the source code of the design which is inefficient especially for a complex SoC with hundreds of IPs), hence increasing the time-to-market. The above situations have raised the need for design understanding approaches which enhance the design experience, debugging and validation facilities, and integrated development environments. The results of design understanding tools not only help designers when (proper) documentation is not available but also can be used for the following tasks.

- Validating that given documentation still corresponds to the current state of the VP implementation.
- Helping designers to understand intricacy of VPs by generating supplementary information from the VPs' implementation and adding them to the existing documentation. This additional information primarily assists designers who have to face hundreds of documentation sheets to understand complex models.
- Facilitating other steps of the design process, such as design space exploration, debugging, validation, verification, or synthesis of the VP model. The predominant way to apply each of the aforementioned tasks to SystemC-based VPs is the simulation-based techniques (mainly due to the object-oriented nature and event-driven simulation semantics of SystemC). To take advantage of these techniques, accessing the detailed structure and behavior of the design is necessarily required. As illustrated in Fig. 1.2, for a given SystemC VP, the design understanding tools can provide designers with a comprehensive and accurate set of information about the design. However, in the absence of a proper tool, this problem is often

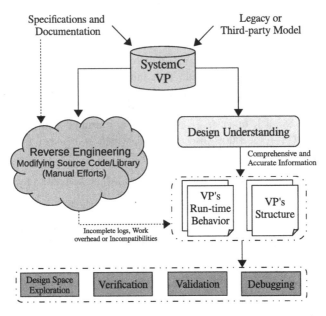

Fig. 1.2 The effect of *Design Understanding* phase on the design process

solved by some reverse engineering, such as manually adding instrumentation codes to either the design or the library. Even if the VP documentation is available (but for legacy or third-party model mostly not), this solution usually results in incomplete logs, work overhead, or incompatibilities.

Therefore, the results of this early analysis step provide designers with an *Intermediate Representation* (IR) of the whole system's behavior and structure (which is based on simulation) and can be served as a starting point for different applications in the design process. Focusing on applications is the second part of this book which pays off the main advantages of a strong design understanding approach.

The primary goal of design understanding approaches is to automatically generate or retrieve more abstract representations of hardware designs and present them in a proper format allowing designers to grasp their essential structures and behavior quickly. The straightforward procedure to access this information is to extract both the design's structure and behavior from the source code using frontend or parser-based approaches. While this technique could be easily used to analyze systems described through classic HDLs (e.g., VHDL or Verilog) where hardware structures are modeled statically, it mostly cannot be applied for the SystemC-based VPs. In the ESL, to cope with design complexity, VPs are usually implemented in a modular way. The module hierarchy of a SystemC VP is dynamically constructed during the execution of the elaboration phase (the beginning of execution time when the modules are created and signals are initialized) of the design. This allows designers to implement a VP based on regular structures using loops and conditional statements. Moreover, this dynamic generation feature enables them to parameterize the design, making the modification, extension, or IP reuse steps significantly easier. Therefore, the design understanding methods, which statically analyze a VP, restrict to a subset of SystemC code, which describes the construction of the model statically.

Dynamic analysis methods have been developed to overcome the limitation of static methods. The basic idea behind the dynamic methods is to extract the information of a design that cannot be retrieved from the source code or at compile-time. Thus, this information is extracted at least at the end of the design's elaboration phase after the module hierarchy is created. These methods are usually implemented as a front-end using the SystemC standard features and API to retrieve the dynamic information, such as the modules hierarchy. While describing the design structure by using dynamic approaches sound well, the main challenge for these methods is the extraction of SystemC design's behavior. They are mostly unable to monitor the behavior of a VP as they do not trace the VP's behavior during the execution time. The information that describes how different modules of the design interact or what the actual value of modules' signals is, needs to be extracted to describe the VP's behavior accurately. However, extracting information about the behavior inside a running C++ program is a non-trivial task.

Hybrid approaches use the best features of the static and dynamic methods to extract the designs' structure. The extracted structural information is linked by using a post-process analysis. To monitor the VP model's behavior, they usually utilize the extracted structural information of the VP to access its run-time behavior. Current hybrid methods, which are able to analyze the SystemC VPs' behavior, have been developed in different ways such as modifying the source code, altering the existing SystemC library or compiler. Overall, the main drawbacks of the existing hybrid methods can be listed as the following.

- *Lack of extracting behavior*: Most of them can only extract a limited set of information that only describes the structure of a given VP model and do not reflect its run-time behavior.
- *Lack of supporting TLM constructs*: They mostly can be applied to a restricted range of SystemC designs (i.e., the VP that is implemented in SystemC cycle-accurate model and has RTL behavior description) and do not support TLM constructs.
- *Intrusiveness*: They mostly rely on manipulating the source code (which is expensive for manual processes), the SystemC library, kernel or interfaces (which may be an issue for the application of several approaches in parallel, future updates, or restrictive environments due to the compatibility problems).

According to the discussions above, there is a need to fill the gap between the traditional SystemC-based VP analysis approaches and the specific requirements for design understanding at the ESL. In this context, this book aims to achieve the following two primary goals.

1. Providing an automated, non-intrusive (or as less intrusion as possible), and comprehensive design understanding methodology to analyze all types (both RTL-based descriptions referring to cycle-accurate model and TLM constructs) of SystemC-based VPs. The proposed methodology must be able to extract different depths of information from a given SystemC VP, describing both structure and its run-time behavior. Moreover, the extracted information must be presented in structured and hierarchical formats, enabling designers to quickly grasp the complexity of VPs.

2. Utilizing the results of the design understanding for other tasks in the design process such as verification, security validation, and design space exploration. These applications are served to demonstrate the advantages of the design understanding methodology and show how the extracted information can be used (off-the-shelf or with minimum translation effort to the desired formats) to facilitate performing the aforementioned tasks.

1.1 Overview

This book explores SystemC-based VP analysis at the ESL to achieve two main objectives, i.e., (1) providing a comprehensive and automated design understanding framework to help designers in the understanding of the VPs' intricacies, and (2) utilizing the results of the design understanding for tasks of verification, security validation, and design space exploration.

The design understanding phase focuses on retrieving both static and run-time information of ESL models and presenting them in a structured format. The extracted information describes both the structure (architecture) of designs (i.e., the modules, attributes, member functions, binding information of the modules' signals, and parameters of functions) and their simulation behavior (i.e., the sequence of the modules' activation, value changes, and function calls during the execution). To illustrate how this behavioral information can be processed to facilitate the understanding of a given VP's intricacy, it is automatically transformed into a *Value Change Dump* (VCD) file and a set of *Unified Modeling Language* (UML) [94] diagrams for SystemC cycle-accurate and TLM-2.0 designs, respectively. The UML diagrams specify the behavior of transactions, allowing designers to use a familiar, abstract design language to understand the behavior that is occurring during the simulation.

In the second phase, to show how the extracted information assists the design process, this information is used for the following tasks: (1) validating that a given SystemC-based VP adheres to the TLM-2.0 rules and its specification, (2) performing security validation of a given VP-based SoC against a set of predefined security rules, and (3) proposing a hybrid analysis on the simulation behavior of a given SystemC-based VP in order to detect which parts of the model are candidates for approximation.

1.2 Outline

This book consists of seven chapters, including the current introductory chapter. Chapter 2 presents the necessary background information for this book. This chapter includes brief introductions on the SystemC language and its TLM framework, and the related works in the design understanding domain. Chapters 3, 4, 5, and 6 present the main contributions of the book. Chapter 3 presents the proposed design understanding methodology, while others show different applications of the design understanding usefulness in the design process. These chapters are briefly described in the following.

- Chapter 3 introduces the design understanding methodology from two perspectives, i.e., debugger-based and compiler-based approaches. In both techniques, the primary goal is to extract both the structure and run-time behavior of a given SystemC VP, post-analyze this information to reduce the complexity of its presentation for designers, and finally, visualize the classified information.
- Chapter 4 presents an application of design understanding for verification of SystemC VPs at the ESL. In this respect, a hybrid automated approach is presented to formally verify the simulation behavior of a given SystemC TLM-2.0 VP against TLM-2.0's rules and the VP's specifications. This chapter also shows how the simulation behavior of the VP is transformed into a set of finite state machines and then checked against TLM-2.0 rules and its specifications by a model checker.
- Chapter 5 illustrates another application of design understanding which is an *Information Flow Tracking* (IFT) approach for security validation of a given VP model at the ESL. The focus of the proposed approach in this chapter is to detect the security violations related to the most occurring threat models; confidentiality and integrity by analyzing the VP's simulation behavior. The precise violation paths are reported back to the verification engineer to either replace the (third-party) VP model or update the security policy.
- Chapter 6 illustrates a use case of design understanding for the task of VPs' design space exploration. The extracted information of a given SystemC VP is used to find the resilient (approximable) portions of the design by using machine learning techniques. The chapter also shows how the simulation behavior of the VP is translated into sets of observation to be used as the input of learning algorithms. Moreover, it presents a bottom-up approximation degree analysis once the resilient portions are identified. The analysis is performed to determine the maximum (1) error rate that each resilient portion can tolerate, (2) number of resilient portions that can be approximated at the same time, and (3) number of modules that can be approximated simultaneously.

This book concludes with Chap. 7, which also briefly discusses the potential research directions for the future work in the field of design understanding and its applications.

Chapter 2
Background

This chapter aims at keeping this book self-contained by providing brief introductions on the basic concepts required for the following chapters. The chapter includes four introductory sections. The first and second sections give a more detailed introduction to the SystemC language and its TLM framework, respectively. The third section introduces the related works in the design understanding domain. This is followed by four sections discussing various information extraction approaches (i.e., static, dynamic, and hybrid) and a summary of their main features. A summary of the key insight of this chapter is presented in Sect. 2.4, explaining the major limitations of the existing design understanding approaches and the need for a promising way to address these limitations.

2.1 SystemC

SystemC is a C++ based system-level design language, which provides an event-driven simulation kernel. It has become a de-facto standard for hardware/software co-simulation and creating VPs at the ESL. The SystemC has been developed by the *Open SystemC Initiative* (OSCI) and is described in IEEE standard 1666–2005. To model a hardware system at the ESL by using a software language (i.e., C++), SystemC provides designers with hardware-oriented constructs which are typically not a part of the C++ language. These constructs include module hierarchy, structure and connectivity, specific hardware data types, time model, communication mechanisms between concurrent units, and concurrency model. The SystemC library consists of classes, macros, and templates which can be used to model a concurrent system using the constructs as mentioned earlier.

A SystemC VP is structured by means of modules. Modules are classes that inherit from the *sc_module* base class. A module includes a part of the system and communicates with other modules through its communication ports. Communica-

© Springer Nature Switzerland AG 2020

M. Goli, R. Drechsler, *Automated Analysis of Virtual Prototypes at the Electronic System Level*, https://doi.org/10.1007/978-3-030-44282-8_2

tion ports can be interconnected by using channels. The concept of channel covers a wide range of hardware complexity from a complex bus down to a simple wire. Apart from the native C++ data type, SystemC includes a wide range of hardware data types that support a various range of bit length from 8-bits to 64-bits for both integral and fixed-point values. Moreover, it supports the non-binary hardware types by using a four-state logic data type $(0, 1, X, Z)$.

To show the notion of time, SystemC uses *sc_time* class, which allows designers to use different time units from seconds down to femto-seconds. For the hardware models that need a clock as the notion of time, it includes a class known as *sc_clock*.

The behavior of a SystemC module is defined by one or more SystemC processes. The SystemC processes are a member function of *sc_module* registered with the SystemC simulation kernel using one of the three available macros, i.e., *SC_THREAD*, *SC_CTHREAD*, or *SC_METHOD*. Each SystemC process has a sensitivity list, its feature, and a use case as the following.

- The *SC_THREAD* process is only executed once by the simulation kernel and can be suspended by calling the SystemC *wait* function. The process state changes from suspend to active when one of the events occurs on its sensitivity list.
- The *SC_CTHREAD* process is a sub-type of *SC_THREAD*, which is only sensitive to the edge of a clock input.
- The *SC_METHOD* process is activated whenever one of the events on its sensitivity list occurs and runs to its ending.

Generally, from the software point of view, SystemC processes are C++ member functions and threads of execution, while from the hardware point of view, they are independent timed circuits that are executed concurrently. A SystemC module can also have a C++ member function to implement a part of its behavior.

The SystemC simulation kernel can be used to execute a SystemC model. As illustrated in Fig. 2.1, it has three main phases, including elaboration, execution, and cleanup.

The SystemC modules' instantiation and initialization are performed by executing their constructors during the elaboration phase. In this phase, the connections between the modules are built. As SystemC modules are instantiated and connected by executing a C++ code, designers can use different valid C++ language

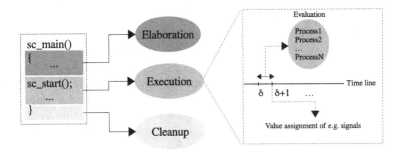

Fig. 2.1 SystemC simulation kernel overview

constructs, e.g., by reading the configuration of the modules from a file or receiving them from command-line arguments. This includes the execution of instructions before the *sc_start()* function call in the *sc_main* function.

In the execution phase, the SystemC kernel provides an illusion of process execution concurrency. It means that the simulation of modules' processes is performed sequentially and all values are assigned at a certain time to the corresponding SystemC channels (e.g., signals). This certain time is considered as a *delta* cycle. Thus, process execution continues until the simulation of the other processes is required to keep behaviors aligned in time. When no additional simulation processes need to be evaluated at that delta cycle, outputs are updated (e.g., value of signals).

This process continues until no further SystemC process needs to be run. This is considered as the cleanup phase where the simulation ends.

2.2 SystemC TLM-2.0

Transaction level modeling was introduced by Synopsis company in 2000, allowing designers to model a system even in a more abstract and faster way. The TLM-2.0 as the current standard is widely used for early system design and verification by supplying reference models for less abstract levels of the design process (such as RTL implementations). TLM introduces two essential concepts over the SystemC foundation, which are *transaction* and *temporal decoupling*.

The transaction concept is the heart of the TLM approach allowing designers to describe a model in terms of abstract communication using the base protocol, standard interfaces (e.g., *b_transport* and *nb_transport*), the global quantum, initiator and target sockets, the generic payload, and the utilities. It enables designers to abstract away the implementation details related to the computation of IPs that may be added at the lower levels of abstraction. Thus, communication (among different IP cores) is the main part of TLM models, which is performed by using abstract operations (i.e., transactions). A transaction is a data structure (i.e., a C++ object) passed through TLM modules using function calls. Thus, data exchange is modeled by transactions. The TLM-2.0 standard suggests that designers use the generic payload class type for transaction objects passed through the communication interfaces as it improves the interoperability between different TLM modules, especially in case of memory-mapped bus models. Moreover, it provides designers with an off-the-shelf structure to transfer information with different member functions to get or set data.

Temporal decoupling comes with the idea of permitting different processes of a SystemC design run ahead in a local time without actually advancing simulation time. This process can continue until they reach the point when they require to synchronize and interact with the other parts of the system. At this point, the process returns control to the simulation kernel and resumes whenever the simulation time catches up with the local time. As this process reduces the scheduling overhead of the simulator, temporal decoupling can result in high-speed simulation even more

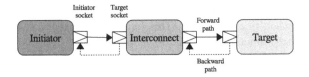

Fig. 2.2 A simple structure of a SystemC TLM-2.0 VP including initiator, interconnect, and target modules

than the traditional event-driven model used by the SystemC simulation kernel. As illustrated in Fig. 2.1, the execution phase of the SystemC simulation kernel relies on a single simulation time value that is advanced for all processes. It means that all processes are stopped and evaluated by the kernel, and required changes are applied to update outputs. TLM allows functions in different modules call each other and invoke the required actions instead of invoking the scheduler to manage the execution of SystemC processes.

As shown in Fig. 2.2, a TLM model may include initiators, interconnects, and targets. An initiator module initiates new transactions through the initiator socket, an interconnect acts as a transaction router, and forwards the incoming transactions without modifying them, and consists of initiator and target sockets. The target module is the end point for the transactions and responds to the incoming transactions. The response can be returned to the initiator in two ways, which are the return path of the transport function in the forward path or the backward path. The latter is performed through an explicit passed transport function call on the opposite path from the target back to the initiator.

Communication between two TLM modules in a VP can be performed based on two-timing models: *loosely timed* (LT) and *approximately timed* (AT). The former is appropriate for the use case of software development, while the latter is necessary for architectural exploration and performance analysis. It is also possible to combine both the aforementioned timing models.

The LT model is implemented using the blocking transport interface (*b_transport*), allowing only two-timing points to be associated with each transaction. The first timing point is the request, while the second is the response. These two-timing points can occur at the same simulation time or at different times. The main advantages of the LT model are the support of temporal decoupling and the completeness of a transaction in a single function call, resulting in very fast simulation speed.

The AT model is implemented using the non-blocking transport interface (*nb_transport*), providing multiple phases and timing points for a transaction. The timing points explicitly mark the transition between phases. Due to the TLM-2.0 base protocol, there are four timing points marking the start and the end of two transaction's phases, which are request and response. These timing points are defined as BEGIN_REQUEST (BRQ), END_REQUEST (ERQ), BEGIN_RESPONSE (BRP), and END_RESPONSE (ERP). It is also possible to define specific protocols with further timing points, which may cause the loss of direct compatibility with the generic payload. However, due to these multiple timing points,

Table 2.1 Different types of the TLM-2.0 base protocol transaction

TM	TT	Communication interface call	Return status	Phase transition
LT	T_0	b_transport	TC	–
	T_1	nb_transport_fw	TC	BRQ
	T_2	nb_transport_fw	TU→TC	BRQ→BRP→ERP
	T_3	nb_transport_fw	TU→TA	BRQ→BRP→ERP
	T_4	nb_transport_fw/nb_transport_bw	TU→TA→TA	BRQ→ERQ→BRP→ERP
	T_5	nb_transport_fw/nb_transport_bw	TU→TC	BRQ→ERQ→BRP
	T_6	nb_transport_fw/nb_transport_bw	TU→TC	BRQ→BRP
AT	T_7	nb_transport_fw/nb_transport_bw	TU→TU	BRQ→ERQ→BRP→ERP
	T_8	nb_transport_fw/nb_transport_bw	TA→TC	BRQ→BRP
	T_9	nb_transport_fw/nb_transport_bw	TA→TA→TC	BRQ→BRP→ERP
	T_{10}	nb_transport_fw/nb_transport_bw	TA→TU	BRQ→BRP→ERP
	T_{11}	nb_transport_fw/nb_transport_bw	TA→TA→TC→TC	BRQ→ERQ→BRP→ERP
	T_{12}	nb_transport_fw/nb_transport_bw	TA→TA→TC	BRQ→ERQ→BRP
	T_{13}	nb_transport_fw/nb_transport_bw	TA→TA→TU	BRQ→BRP→ERP

TM Timing model, *TT* Transaction type, *TC* TLM_COMPLETED, *TA* TLM_ACCEPTED, *TU* TLM_UPDATED, *BRQ* BEGIN_REQUEST, *BRP* BEGIN_RESPONSE, *ERQ* END_REQUEST, *ERP* END_RESPONSE

the possible performance gain, that can be achieved by the temporal decoupling concept, is nullified in the AT model. Due to the combination of these phases and timing points, 13 unique ways of transaction type are defined in the base protocol.

In summary, Table 2.1 shows different types of the TLM-2.0 base protocol transactions and describes them based on the communication interface call, return status of the interface call and the transaction's phase transitions.

2.3 Related Works

Analyzing SystemC-based VPs (e.g., for design understanding) is an active field of research. Several methods have been developed to achieve this goal, each of them with its features and issues. These methods can be divided into three main categories based on whether they rely on static, dynamic, or hybrid techniques. In the following, each type of the aforementioned design understanding techniques is discussed in detail.

2.3.1 Static Methods

Static approaches rely on extracting information from the source code or its compiled binary model using parsers [29, 44, 59, 88, 103] or the existing C++ front-

Fig. 2.3 General
methodology of the static
design understanding
methods

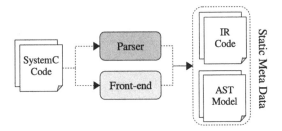

ends [8, 10, 60]. They do not (by definition) analyze the execution of the models. Their results can only describe some information related to the structure of a model and in the best case, the result can be represented in an *Abstract Syntax Tree* (AST). The general flow of the static approaches is illustrated in Fig. 2.3.

The C++/SystemC parser ParSyC [29] is based on the *Purdue Compiler Construction Tool Set* (PCCTS) [87]. It translates a given SystemC model into an AST format. The generated AST is converted to an intermediate representation for semantic consistency and finally synthesized to a netlist. As the final result of the approach is to generate a netlist form of the SystemC model, it can only be applied to the synthesizable version of SystemC designs. This feature limits the parser only to support a subset of SystemC models.

The *Karlsruhe SystemC Parser Suite* (KaSCPar) [59] is based on the *Java Compiler Compiler* (JavaCC) and preprocessor JJTree. Same as ParSyC, it describes both the structure and behavior of a given SystemC model in terms of AST and presents it in XML format. To generate this presentation, KaSCPar parses the SystemC keywords in a given SystemC model by tokenizing the structural and behavioral features of the SystemC language. However, the approach does not support the whole C++ standard. Thus, compiler-specific additions to the standard library are not usually supported.

SystemCXML [8] uses an XML-based approach relying on Doxygen [117] to extract structural information from SystemC designs. The analysis is performed in three main phases. First, the source code of a given SystemC design is parsed by Doxygen in order to generate an XML output including all design's structural information. This information is embedded with XML tags. Second, the generated XML model is again parsed using the Xerces-C++ XML parser [31] to create an *abstract system-level description* (ASLD) XML file. Finally, the generated XML model from the previous phase is analyzed to generate an IR model of the design. Thus, it provides designers with a source-to-source translation of a SystemC design's structure to an IR model. However, SystemCXML is limited to extract behavioral information. Moreover, it does not support TLM constructs.

SystemPerl [103] takes advantage of regular expressions and a set of Perl scripts to parse SystemC designs and access various SystemC constructs. However, to accurately analyze a given SystemC model, SystemPerl needs designers to modify the source code and provide hints in the program. Essentially, this enables the approach to identify the required constructs that must be extracted, a process that may require a lot of programming effort. Moreover, this approach (same as SystemCXML) only extracts structural information from a given SystemC design and not the behavioral data. Both aforementioned drawbacks limit the use case of SystemPerl for the task of design understanding.

Scoot [10] utilizes a C++ front-end to translate the SystemC designs' source code into a *control flow graph* (CFG). Afterward, the generated CFG is statically analyzed to generate a formal representation of design's structure (module hierarchy, sensitivity list of the processes, and binding information of modules' ports), and a C++ model of the source code. However, for this analysis, Scoot requires designers to modify the standard scheduler of the SystemC kernel and the SystemC header files. These two drawbacks restrict the utilization of Scoot for the task of design understanding as an existing project may need some burdensome revising process of the given source code, which might not always be possible.

systemc-clang [60] use the Clang compiler to extract the structure and behavior of a given SystemC model. It works on the AST of a given SystemC design, which is generated by the compiler. The approach is able to support both the SystemC cycle-accurate model and the TLM constructs. However, as the whole analysis is performed on the AST of the design, it inherits the same limitation of other static approaches. This means that the extracted static information in this case only describes the design's structure and limited semantics of its behavior.

Recently, an automated approach was introduced to automatically retrieve the structure and behavior of a given ESL model [98]. It uses the ROSE compiler to generate an AST model of the design, which is then used to generate an IR. The IR is analyzed to extract both the architectural information of the model and the behavior as graphical multi-thread communication charts. However, solely relying on a static analysis, the method shares the limitations of previous static approaches. It cannot consider parameters which are set at run time (and may affect the design's behavior). The behavioral information is restricted to only describe high-level interaction of modules—thus, behavior such as value changes of a module's ports and function's variables during execution are not traceable. Moreover, the method does not support designs, including pointers or array indices in port mappings.

2.3.2 Dynamic Methods

Since the extraction of dynamic information is necessary to describe both the structure and behavior of a given ESL design, the dynamic methods have been developed. Besides extracting the static data, dynamic approaches retrieve additional information of a given SystemC design after starting its execution.

Quiny [99] is a SystemC front-end, which uses a dynamic method to retrieve the hierarchical and the behavioral information from the SystemC model. The first step of the method is to modify the SystemC library and generate the so-called *Quincy* library. Then, the SystemC model is compiled and linked to the Quincy library. Subsequently, when the model is executed, Quincy produces an IR of the

SystemC code. It means that the method translates all SystemC types, operators, and statements into a C++ code presenting the IR of the SystemC model. To handle some specific data type such as *unsigned int*, pointers, and declarations, a manual effort is required. For example, the type *Q_UINT* must be used by designers instead of the C++ type *unsigned int*. These features dramatically restrict the automation feature of the method. Moreover, modifying the SystemC library may cause a compatibility issue for other applications or approaches in parallel (an intrusive solution that comes at a price).

2.3.3 Hybrid Methods

The main challenge for the SystemC dynamic analysis methods is the retrieval of the behavior of the SystemC VP. To overcome this limitation, hybrid methods have received the most attention lately. They take advantage of the best features of the static and dynamic methods and combine them to trace the run-time behavior of a VP model. In this section, we give an overview of hybrid methods based on whether they support TLM constructs or only analyze cycle-accurate SystemC VP, illustrating their features and issues.

2.3.3.1 Methods that Do Not Support TLM

The hybrid analysis methods of SystemC VPs that do not support TLM constructs are introduced in this section. These approaches extract either the model structure or dynamic behavior of the model, or both.

In [34], the AST of a SystemC model is retrieved by parsing the model using a PCCTS-based parser. To extract dynamic information, an instrumented version of the source code is generated by adding some *recorder function*. The state of all variables of the model is recorded by executing its elaboration phase. Using the PCCTS-based parser limits the available SystemC constructs as it does not fully support the entire instruction set of C++.

PinaVM [74] extracts the structure of a SystemC model from the translated version of the source code into LLVM [65] bit-code by executing its elaboration phase. To extract the dynamic information, it specifies the parts of the source code which contain the parameters of interest (e.g., the address of ports or events in SystemC constructs). Afterward, new functions are constructed to be added to the model during its compilation. Those parameters are retrieved using the generated functions during the model's execution. PinaVM takes advantage of the LLVM project to analyze SystemC models, limiting it to setups that are built using LLVM.

SHaBE [12] retrieves the static data by utilizing the *GNU DeBugger* (GDB) [105] and extracts the dynamic information using a GCC plugin. In the next step, the dynamic information is linked with the hierarchical information and stored as an intermediate representation. The method has limitations to extract some static and dynamic information of SystemC constructs (e.g., SystemC primitive channels or processes sensitive to particular events).

The method presented in [106] uses *Aspect-Oriented Programming* (AOP) to extract the behavioral data. AOP is a paradigm that allows the designers to write re-factoring rules that are applied before compiling a program (a process called weaving). This approach comes with several pitfalls. Debugging AOP setups is a complex task, just like setting up a working AOP environment. Furthermore, the current implementation of AspectC++, which can be used in tandem with SystemC, does not support, e.g., join points for field access (i.e., field variable assignments cannot be tracked), privileged aspects, templates, or macros, which limits the goal of arbitrary behavior tracing.

2.3.3.2   Methods that Support TLM

The hybrid methods that can analyze TLM models are introduced in this section. These approaches extract either the model structure or transaction behavior, or both.

Pinapa [80] retrieves the information of SystemC models in two steps. First, the AST of the models is extracted using a C++ front-end. Second, the elaboration phase of the models is executed to extract their dynamic information. The extracted information is linked to the AST to create the final result. Although the method describes the structure of a given SystemC model, it does not provide any information to reflect the design's behavior, such as the order of function calls or processes activation.

The *SystemC Verification* (SCV) library by the *Open SystemC Initiative* (OSCI) provides designers with a set of APIs to record transactions into a database. The APIs are divided into three different transaction collection classes that can be instantiated during the execution of a SystemC TLM-2.0 model. The results obtained by this method can be analyzed by some commercial tools such as Cadence Incisive [13] or Novas Verdi [85] with modifications to the result-file format. SCV introduces some overhead in execution time. The method is an intrusive solution to extract the behavior of a SystemC TLM-2.0 model as the source code needs to be manipulated. For an intricate design, this manual process is a non-trivial task.

DUST [63] is a SystemC TLM-2.0 analysis framework that extracts both structural and behavioral information of a TLM-2.0 model. It retrieves the model's hierarchy at the end of the execution of its elaboration phase and presents it in an XML format. It describes the behavior of the model by recording transactions at run time. DUST works by enhancing some SystemC objects (e.g., *sc_port* to *sc_dust_port*) and using SCV constructs which are added to the source code. Due to this intrusive solution, its application may thus be limited if the SystemC library is updated, but the given framework is not. Additionally, existing sources need to be updated to use DUST's types instead of the standard SystemC types—a solution that may require considerable work.

The method presented in [107] takes advantage of debug symbols to extract static information and SystemC API calls to retrieve dynamic data during the execution of a SystemC model. The dynamic information extracted by this method only reflects the structure of the model. The simulation behavior is not captured at all, though, leaving designers with static model descriptions. Moreover, the method only supports the debug symbols that are generated by Microsoft VC++ and not GCC or Clang-LLVM [65].

2.3.4 Summary

The results of studying different SystemC VP analysis approaches are summarized in Table 2.2. This table shows five essential features that a SystemC VP analysis approach must include to enable designers to have a comprehensive analysis on various aspects of the VP. As illustrated in Table 2.2, the static aspect of structural information was well studied by the previous solutions. However, the following limitations still exist.

1. Low degree of automation,
2. Custom code annotations or language constructs,
3. Lack of extracting VPs' run-time behavior, and
4. Lack of supporting TLM constructs.

Concerning the first and second limitations: the existing approaches mostly require designers to manually manipulate the source code (which is an expensive process even for simple design), the SystemC library and interfaces, or the SystemC kernel (which may be an issue for the application of several approaches in parallel, future updates, or restrictive environments). Generally, requiring manual effort by designers is an error-prone process, increases the setup time of using the approach, and reduces the degree of automation. Moreover, modifying the standard workflow (i.e., SystemC library, interfaces, or kernel) may affect the functionality or timing behavior of the design. Thus, the obtained results may not be identical to the original one.

Concerning the third and fourth limitations: although some of the existing approaches are able to describe the static aspect of a given VP's behavior (e.g., the number of modules' processes and their relationship), the run-time behavior (e.g., order of modules activation, and run-time value of transactions or variables) cannot be extracted. It means that most of them extract a limited set of information that only describes the structure of the model and does not reflect its run-time behavior. The other major limitation is that most of them can only be applied to a restricted range of SystemC designs (e.g., only synthesizable SystemC VPs) and do not support TLM constructs. Those approaches that can support TLM designs do not provide designers with a suitable form presentation of designs' behavior. The generated results mostly are in the XML formatted files. The XML model can be useful for VP's structure presentation as the amount of information to be presented is constant and does not change during the simulation. In the case of

Table 2.2 Summary of the SystemC-based VP analysis approaches' features

Analysis type	Method	Automated	Non-intrusive	Structural information		Behavioral information		TLM-2.0
				Static	Dynamic	Static	Run-time	
Static	ParSyC [29]	✓	✓	✓	✗	✗	✗	✗
	KaSCPar [59]	✓	✓	✓	✗	≈	✗	✗
	SystemCXML [8]	✓	✓	✓	✗	✗	✗	✗
	SystemPerl [103]	≈	✗	✓	✗	✗	✗	✗
	Scoot [10]	≈	✗	✓	✗	✗	✗	✗
	systemc-clang [60]	✓	✓	✓	✗	✓	✗	✓
	[98]	✓	✓	✓	✗	✓	✗	≈
Dynamic	Quiny [99]	≈	✗	≈	✓	✓	✗	≈
Hybrid	[34]	✓	✗	✓	≈	✓	✗	✗
	Pinapa [80]	✓	✓	✓	✓	≈	✗	✓
	PinaVM [74]	✓	✓	✓	✓	≈	≈	✗
	SHaBE [12]	✓	✓	✓	✓	≈	✗	✗
	[106]	≈	✓	✓	≈	≈	✗	✓
	DUST [63]	≈	✗	✓	✓	≈	≈	✓
	[107]	✓	✓	✓	✓	≈	✗	✓

✓ support, ≈ not fully support, ✗ not support

behavior presentation, the XML file can grow very large (depending on the running software or application), which is not easy to be understood by designers.

2.4 Conclusion

In this chapter, first, an introduction to the SystemC and its TLM framework was presented. Second, the existing SystemC design understanding approaches were studied in detail. The results of this study showed that the existing solutions have two major limitations in terms of precise behavior extraction of a given ESL model and custom code annotations or language constructs. The first limitation is that most of them extract a limited set of information that only describes the structure of the model and does not reflect its run-time behavior. The second limitation is that most of them can only be applied to a restricted range of SystemC designs and do not support TLM constructs. Those approaches that analyze the SystemC TLM-2.0 models mostly require designers to manually manipulate the source code (which is an expensive process even for simple design) or the standard workflow (which may be an issue for the application of several approaches in parallel, future updates, or restrictive environments).

Therefore, we conclude that for a given SystemC-based VP currently no automated analysis approach is available (without having the issues mentioned above) that can extract both structural information (static and dynamic parameters which need to describe, e.g., its modules hierarchy) and its run-time behavior. Hence, to overcome the aforementioned drawbacks and enable designers to have complete access to both structure and simulation behavior of VPs, in the following chapter, two SystemC VPs analysis approaches are presented.

Chapter 3
Design Understanding Methodology

Analyzing a given SystemC-based VP is an intricate task. As discussed in Chap. 2, this is mainly because of various complex constructs, concurrency model, and protocol rules defined in SystemC language, enabling designers to model VPs at the ESL. Designing a complex system may require revision, modification, or extension that must be applied to the existing IPs of the VP. These processes become more complicated when designers need to use some third-party IPs (that usually do not have proper documentation) to reduce time-to-market constraint. However, before applying any changes or modifications to a given VP, designers need to know what the VP's structure is and how different IP cores behave and communicate with each other. We reported in Chap. 2 that existing SystemC VPs' analysis approaches fail in providing designers with a fully automated solution, which extracts the required information from a given VP to properly understand both its structure and behavior.

In this chapter, first, to demonstrate the necessity of the SystemC-based VPs analysis and visualization (design understanding) in the design process, two motivating examples are presented. Next, the required information that must be extracted from VPs for design understanding purpose at the ESL is introduced. The SystemC VPs analysis methodology is introduced in two perspectives, i.e., debugger-based and compiler-based approaches. Then, the visualization of the extracted information is described. Finally, the experimental results related to each approach are presented.

3.1 Motivating Examples

The examples in this section show that the need for design understanding approaches in the design process is essential. The first example illustrates that the extension or revision of a VP implemented in SystemC TLM-2.0 can be

© Springer Nature Switzerland AG 2020
M. Goli, R. Drechsler, *Automated Analysis of Virtual Prototypes at the Electronic System Level*, https://doi.org/10.1007/978-3-030-44282-8_3

more straightforward with the help of an automated analysis approach, which provides detailed information about both structure and behavior (e.g., types of a transaction) of the VP. In the second example, we show that the debugging task in the design process for a VP implemented based on the cycle-accurate model in SystemC can be easier if designers could access the precise run-time behavior of the VP.

3.1.1 Example 1

Consider the third-party *LT_AT_BUS* VP (inspired by Aynsley [6]) shown in Fig. 3.1 that the documentation is not available (or poorly written). The VP includes six modules, which differentiate based on underlying base protocol transactions: two initiators (*Initiator_A*, *Initiator_B*), one interconnect (*LT_AT_BUS*), and three targets (*Memory_A*, *Memory_B*, *Memory_C*). The *Initiator_A* module communicates with target modules through *LT_AT_BUS* by generating four types of AT transactions. Two types T_0 and T_4 to access *Memory_A* (each type for different memory address ranges), and types T_1 and T_2 to access *Memory_B* and *Memory_C*, respectively. The *Initiator_B* module only generates transactions of type T_3 to communicate with all target modules. For example, consider the communication between *Initiator_A* and *Memory_A* (the gray components in Fig. 3.1). The *Initiator_A* module generates transaction types T_0 and T_4 to access memory address range ($0x00$ to $0x0A$) and ($0x0B$ to $0xFF$) of the *Memory_A* module based on the functions call and timing phases described in Table 2.1, respectively. Now consider a scenario that may happen during the design process.

Designers decide to reuse or revise some modules of the VP. For example, they want to modify the AT base protocol transaction type T_4 of the *Initiator_A* module and change it to T_1, including fewer transition phases to gain performance. This modification also needs to be applied to the *LT_AT_BUS* and *Memory_A* to properly build a communication path between the initiator and target modules through the interconnect. Before any changes can be applied to the VP, it needs to be adequately understood. However, lacking a (proper) documentation makes the understanding process very complicated.

Fig. 3.1 The architecture of the *LT_AT_BUS* VP

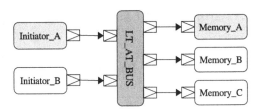

3.1.2 Example 2

As the second example of the design understanding use case, consider the *2-stage pipe* VP (Fig. 3.2) implemented in SystemC. The standard SystemC design example in [1] inspires the VP; however, it provides other functionality to support our motivating example. The design includes two modules *M1* and *M2*, and performs a set of algebraic operations to generate the final results (*out1* and *out2* of module *M2*) in two steps.

Here we also assume that the only available reference for designers is the value of final output (as reference result) for a specific benchmark (Fig. 3.4), which can be used to validate and debug the design. Moreover, from the VP's specifications, we know the following rules:

- the output port *out3* of module *M1* must have positive values and
- if the input signals are assigned at clock *T*, the final result *out2* of module *M2* is generated at *T+1*.

Now consider the scenario that designers implemented the definition of local variable *temp3* of function *func1* of module *M1* incorrectly (line 30, Fig. 3.2). After executing the design, they found out that at simulation time 9 ns (Fig. 3.3, line 3) the value of final output *out1* of module *M2* is 5.12, which is against the expected output that was given as the reference results (Fig. 3.4, line 3). Since this type of fault is not related to the C++ or SystemC syntax and is a semantic fault, the C++ compiler cannot detect it. Moreover, as the reference model is not available for the design and only a basic specification of the design exists, monitoring the simulation behavior would be the straightforward way.

However, this requires to access the detailed behavior of the design, including the run-time values of local and global variables. The reason is that designers can only see the incorrect final output of the design (i.e., *out1* of module *M2*) and have no information about how other variables are associated to generate the final result. Indicating the relation of variables (local or global) within the design with the faulty output and visualizing it in a proper format could help designers to find the source of the fault. This visualization empowers designers to trace back the value of faulty output to the place where, for the first time, the fault is occurred and propagated to the corresponding final output. The main difficulty here is that as variable *temp3* is locally defined in function *func1* of module *M1*, the conventional visualization methods, such as SystemC API trace, do not support tracing the value changes of this local variable.

Please note that the scenario above can be more complicated if conditions in the VP's specification are related to dynamic variables that are defined during execution.

```
1   struct M1 : sc_module {
2    sc_in <bool> clk, ctlM1;
3    sc_in <double> in1, in2;
4    sc_out <double> out1, out2, out3;
5    double genA, genB;
6    void func1();
7   SC_CTOR( M1 ) {
8    SC_METHOD( func1 );
9    dont_initialize();
10    sensitive << clk.pos();};
11   };
12   //------------------------------//
13   struct M2 : sc_module {
14    sc_in<double> in1, in2, in3;
15    sc_in<bool> clk, ctlM2;
16    sc_out<double>  out1, out2;
17    void addsub();
18    double mult();
19    ...};
20   //------------------------------//
21   void M1::func1(){
22    double temp1, temp2, temp3;
23    if (genA > genB){
24     temp1 = in1.read()*genA + in2.read()*genB;
25     ctlM1.write(0);}
26    else{
27     temp1 = in1.read() * genA;
28     ctlM1.write(1);}
29    temp2 =  genA * genB;
30    temp3 =  genA + in2.read()*genB; // buggy instruction -> (
           temp3 = genA-in2.read()*genB)
31    out1.write(temp2);
32    out2.write(temp2 * temp1);
33    out3.write(temp3);}
34   //------------------------------//
35   void M2::addsub(){
36    double tp1, tp2;
37    tp1 = in1.read()*in3.read();
38    tp2 = in1.read()+in2.read()*in2.read();
39    switch (ctlM2.read()){
40    case 0:
41     out1.write(mult() - tp1);
42     break;
43    case 1:
44     out1.write(mult() + tp1);
45     break;}
46    out2.write(tp2);}
47   //------------------------------//
48   double M2::mult(){
49    return (in1.read() * in3.read());}
```

Fig. 3.2 A part of the source code of the *2-stage pipe* VP model

Fig. 3.3 *2-stage pipe* VP's
wrong results of *out1* of *M2*

```
1     3  ns :  address  ->  0
2     6  ns :  address  ->  11.73
3     9  ns :  address  ->  5.12
4    12  ns :  address  ->  7.20
5    15  ns :  address  ->  8.78
6    18  ns :  address  ->  1.43
7    23  ns :  address  ->  3.59
8    28  ns :  address  ->  6.33
9    33  ns :  address  ->  9.98
10     . . .
11   118  ns :  address  ->  10.92
```

Fig. 3.4 *2-stage pipe* VP's
reference results of *out1* of
M2

```
1     3  ns :  address  ->  0
2     6  ns :  address  ->  11.73
3     9  ns :  address  ->  15.52
4    12  ns :  address  ->  19.32
5    15  ns :  address  ->  29.39
6    18  ns :  address  ->  32.69
7    23  ns :  address  ->  39.07
8    28  ns :  address  ->  49.10
9    33  ns :  address  ->  60.33
10     . . .
11   118  ns :  address  ->  78.04
```

3.2 Structural Information

Generally, hardware models include hierarchy to reduce complexity. The architecture of a SystemC VP is a set of IP cores (defined as modules) and connections. In a VP design, IPs are encapsulated as "modules" which are classes that inherit from the *sc_module* base class of the SystemC library. Modules may contain other modules, processes, channels and ports for connectivity. Thus, for a given SystemC VP, the structural information refers to the data that is described in the VP's source code consisting of class and function names, variable information (e.g., module ports, local variables of functions), class hierarchy information, data types, binding information. In the case of TLM VP, in addition to the data above, the structural information also refers to the modules' types (initiator, interconnect, or target), binding information of modules' sockets, and transaction objects and their related variables such as timing annotation, phase and return status of transporting interfaces.

A large part of this information is static data, which can be accessed before the execution time of VPs (e.g., in compile-time) including

- the root name and type of each module (in the case of TLM constructs),
- the name and type of each function,
- the variables of each module, and
- local variables of each function.

However, some part of this information may be identified after the model's execution, which is considered as dynamic data. Examples of dynamic data that may be identified at run-time and cannot be extracted statically from the source code are

- the instance name of each module,
- the binding information of signals and sockets, and
- dynamic variables and parameters.

Therefore, the analysis approach must be able to extract both static and dynamic types of design to describe its structure accurately.

3.3 Behavioral Information

For VPs implemented using SystemC TLM-2.0, the VPs' behavior refers to the run-time information of abstract communication among different IP cores. The main reason is that the TLM paradigm allows designers to abstract away the implementation details related to the computation of IP cores and only focus on communication. Thus, communication is the central part of the VP's behavior. As communication is performed using the transaction concept, describing the TLM VP's behavior is connected to identify the behavior of its transactions. This identification requires to access three essential elements of transactions during the execution time, which are *flow*, *data*, and *type*.

The transaction's flow represents the order of TLM modules taking part in the transaction's lifetime (i.e., the period between transaction construction and destruction). For example, the transaction's flow for both types of T_1 and T_2 generated by *Initiator_A* to access data in *Memory_A* is based on the sequence order $(1) \rightarrow (2) \rightarrow (3) \rightarrow (2) \rightarrow (1) \rightarrow (2) \rightarrow (3) \rightarrow (2) \rightarrow (1)$ where (1), (2), and (3) are *Initiator_A*, *LT_AT_BUS*, and *Memory_A*, respectively. The required information that must be extracted to describe the transaction's flow properly is

- the sequence number of objects' activation,
- the root name of each module taking part in the transaction,
- the role of each module taking part in the transaction (for TLM modules can be initiator, interconnect, or target and for others is global),
- the instance name of each module,
- the name of the current function, its arguments' values and its return value (if available),
- source code information (i.e., line of code and source file name),
- the simulation time, and
- the transaction reference address.

The transaction data denotes the transaction's attributes such as

- data value,
- address,
- command,

- data length, and
- response status.

The transaction's type refers to the transaction's timing model (LT or AT). As mentioned earlier in Chap. 2, TLM-2.0 comes with one type of LT transaction and 13 types of AT transaction. Thus, in the case of the AT model, it requires to be specified which type of base protocol transactions is used. However, this element has not been considered by the existing SystemC analysis approaches [63, 98] as they only extract the transactions' flow (but not their type). The two transaction types T_1 and T_2 have the same communication pattern (flow), but their timing phases are different, these methods fail to distinguish between them. Therefore, to properly understand a given TLM VP, the proposed methodology must be able to extract all the aforementioned elements, distinguish the transactions' unique types and flows and finally present them in a proper graphical format (e.g., UML diagrams), thus designers can quickly grasp the VP's behavior.

In the case of SystemC VP implemented using cycle-accurate model, the VP's behavior is related to its modules' and global functions' behavior. Each module has a behavior defined by one or more processes (e.g., methods and threads) and communicates with other modules through ports, interfaces, and channels (e.g., signals). Moreover, a part of the module's behavior may be defined using local functions. Therefore, to understand the VP's behavior, accessing different states of the VP's entities during its execution time is necessary required. This access includes extracting all signals and variables (local or global) that have been used to implement modules behavior within a process, global, or local functions. In addition, understanding how different modules interact and are activated during the execution time is another essential part of the VPs' behavior. Thus, the order of modules and global functions' activation must be extracted as well. This also must be applied to the lower hierarchy, which is the member functions (i.e., local functions or processes) of a module. Overall, the proposed methodology must be able to extract the following run-time information to describe VP's behavior accurately:

- the simulation time stamp,
- all variables' value of modules and local variables of functions,
- the name of each port,
- binding information of each port,
- the instance name of modules, and
- clock information (if available).

3.4 Information Extraction Methodology

Several techniques can be used as the underlying framework to extract both structural and behavioral information. They are divided into three main categories, which are based on the debugger, compiler, and AOP.

The straightforward way to access the designs' information is the utilization of debuggers. Mainly because of debuggers, e.g., the *GNU Debugger* (GDB) [105] allow designers to monitor the behavior of design during the execution of a model from IPs activation down to single instructions. This is usually performed by stopping the program at specific points and stepping through the execution of certain statements. A debugger is usually fit to help designers in understanding the complexity of a given VP or finding the possible location of bugs in the VP. One notable advantage of using debuggers to analyze SystemC-based VPs is that debuggers, e.g., GDB do not require the availability of the VPs' source code as they can directly work on the executable version of the design. In the case of 3PIP, this option is available for vendors to keep the design techniques private and secure (for their own company) from others and sell only the executable binary of IP with documentation that describes how the design works and it can be integrated with other designs. Thus, using debuggers is the only way to analyze 3PIP models of a given VP without the availability of source code if the debugging information is included in the VP (which is needed and commonly provided by the vendors). However, the whole analysis process usually needs to be performed manually by designers. Therefore, the main challenge to use debuggers as the underlying framework is how to make the whole analysis process automatic.

The second possible technique is the utilization of compilers as the underlying structure to access both the structural and behavioral information of a given VP. Among several open-source C++ compiler infrastructures that exist, GCC and Clang-LLVM are the most popular one and widely used in the academia and industry. For static analysis purpose, which is the first step of accessing the structural information of design, Clang is preferred over GCC due to the following two main reasons.

1. GCC compiler removes the unnecessary information and implicitly simplifies source code during compilation, this may cause problems for the source code analysis approach that builds as a front-end on top of the GCC. Although modifying compilation workflow can be a solution, this comes with several drawbacks as discussed in Chap. 2.
2. The IR of the source code of a program generated by GCC is hard to be supervised for further analysis step, as the generated IR is just a dump of the compiler memory image which does not have a structured format.

Instead, the Clang compiler (which is a front-end for LLVM project) provides designers with analyzing the AST of a program (which is user-friendly and understandable) including the static information in a hierarchical format. Moreover, the high-level C++ interface of the Clang (LibTooling [111]) enables designers to traverse the generated AST of the program and access all information about the program just like its available in the source code but in a structured and well-formed format. Therefore, it is possible to access the structural information of a given VP by analyzing its AST. The other interesting point of the Clang interface is the code generation feature. It enables designers to generate an instrumented version of the VP's source code by inserting new lines of code at any arbitrary points while analyzing the AST of its source code. Hence, the main challenges to using this technique are how to adopt Clang to (1) analyze SystemC constructs to access a given VP's structure and (2) automatically generate an instrumented version of the VP, including the required instruction to properly extract the run-time information that describes its behavior.

The AOP solution is another alternative to analyze SystemC VPs. It allows designers to instrument the source code without altering the original code base, relying on automatic re-writing in an additional step that is executed automatically before compilation. However, when being applied in practice, setting up a working AOP environment for existing, more extensive projects, can quickly turn out a significant amount of work, with the aspect weaver first being required to parse the whole project (which is primarily an issue as soon as libraries are involved) and the resulting code often causing new issues for a given compiler setup. Thus, two major drawbacks of AOP are as the following. First, debugging AOP setups is a complex task, just like setting up a working AOP environment. Second, AOP does not support the description of arbitrary points defining the code that needs to be altered. For example, the current implementation of AspectC++, which can be used in tandem with SystemC, does not support, e.g., join points for field access (i.e., field variable assignments cannot be tracked), privileged aspects, templates, or macros, which limits the goal of arbitrary behavior tracing (as discussed in Chap. 2).

Since both the debugger- (i.e., GDB) and compiler- (i.e., Clang) based approaches provide designers with a readily available means to analyze a program, they are used as the underlying structure for our proposed SystemC-based VPs analysis approaches [37, 40, 41].

3.4.1 Debugger-Based Approach

As illustrated in Fig. 3.5, the core idea of the proposed approach to extract the structure and simulation behavior of a given SystemC-based VP consists of two main phases.

Fig. 3.5 The proposed debugger-based information extraction approach overview

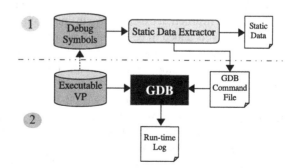

1. The static information of the compiled model is retrieved by analyzing its debug
   symbols to achieve two goals:

   (a) identifying all components and their attributes and member functions which
       are required to describe the structure of the model (e.g., the name and type
       of the modules' variables) and

   (b) automatically generating a set of GDB instructions tailored to be used in
       the next step to extract the model's structure (dynamic data) and trace its
       behavior.

2. The model is executed under the control of GDB using the previously generated
   instructions. The model's structure is retrieved when the execution reaches the
   objects for which the corresponding instructions to extract their information
   were generated. The execution of the model is paused at certain events (such
   as function calls) to record the run-time information.

3.4.1.1   Static Information Retrieval

The model's static information is retrieved by analyzing its debug symbols (which
is generated by the GNU debugger) to extract all SystemC (e.g., name of modules
and their member functions and attributes) and TLM (e.g., the generic payload
data type, initiator and target sockets, and the TLM utilities) constructs. This
information is required to reflect the model's structure and trace its run-time
behavior, e.g., variables value or transactions data in the case of SystemC cycle-
accurate or SystemC TLM-2.0 VPs, respectively.

The retrieved information is translated into a more manageable data format in
which each module is described in a hierarchical structure based on its member
functions and attributes. Unlike other static approaches that consider this type
of information, as a result, this information is used as the foundation to extract
additional run-time information. A *GDB Command File* (GCF), which is used to
program GDB, is automatically generated based on this static data. It controls
the execution of the VP executable model running on GDB to extract the desired
information in the simulation run. This is performed using the breakpoint feature of
GDB. A breakpoint stops the program execution whenever a certain point (e.g., a
function or an instruction execution) in the program is reached. For each breakpoint,
a command list is added to be executed when the program stops at that breakpoint.
The breakpoints are set with the *break* command and its variants to specify the place
where the program is stopped by line number, function name, or exact address in the
program.

Using the motivating examples, we show which type of static information the
debug symbols contains and how it can be used to generate the GCF script.

In the case of SystemC TLM-2.0 design, consider the *LT_AT_BUS* VP. A
part of the static information related to the design's structure is presented in its
generated debug symbols (Fig. 3.6, lines 8–15). It shows that the design contains
a module *Initiator_A* with (among others) an initiator socket *socket* and a member

```
 1   Symtab for file LT_AT_BUS.cpp
 2   ...
 3   Blockvector:
 4   ...
 5   block #169, object at 0x370e870 under 0x39d6930, 4\
 6   syms/buckets in 0x407176..0x40755c, function\
 7   Initiator_A::nb_transport_bw (tlm::tlm_generic_payload&,...)
 8     struct Initiator_A : public sc_core::sc_module {
 9     tlm_utils::simple_initiator_socket<Initiator_A\
10     ,32u, tlm::tlm_base_protocol_types> socket;
11     ...
12     public:
13       virtual tlm::tlm_sync_enum nb_transport_bw (...);
14       ...
15     } * const this; computed at runtime
16     class tlm::tlm_generic_payload {
17       ...
18     } &trans; computed at runtime
19     class tlm::tlm_phase {
20       ...
21     } &phase; computed at runtime
22   ...
```

Fig. 3.6 A part of the debug symbol of the *LT_AT_BUS* design generated by GDB

function *nb_transport_bw* with return type *tlm_sync_enum*. For example, to extract all transactions related to the *LT_AT_BUS* initiator module *Initiator_A*, we need to trace all functions of the module in which a transaction object is referenced. This is performed by finding blocks that contain the information of the module's functions from the debug symbols. The functions' input arguments and local variables are also in the block (Fig. 3.6, lines 16–21). Based on the extracted data, the *nb_transport_bw* function of the module must be traced as well as the transaction object *trans* (Fig. 3.6, line 18), which is its input argument. To trace the function, a breakpoint is set at the beginning of the function body (line 4, Fig. 3.7). For this breakpoint, a set of commands (Fig. 3.7, lines 5–10) is defined that is executed whenever the breakpoint is triggered at run-time. The information that needs to be extracted can be defined within these commands and by default contains information such as the instance name of the module (Fig. 3.7, lines 6 and 7) and the transaction phase (Fig. 3.7, lines 24 and 25).

Regarding the SystemC cycle-accurate model, a part of the generated debug symbol of the *2-stage pipe* VP is presented in Fig. 3.8 (lines 1–17). It illustrates that the VP includes a module called *M1* (line 6) with three input signals *in1*, *in2*, and *clk*, and two output signals *out1* and *out2* (lines 7–12). Moreover, the *M1* module has a member function *func1* with a local variable *temp3* (lines 14–17). For example, Fig. 3.9 (lines 1–13) shows a part of the generated GCF of the *2-stage pipe* VP. Assume that the goal is to extract the value of the local variable *temp3* after the execution of each line of the *func1* function. First, a breakpoint is set based on name of function *func1* (line 1, Fig. 3.9) to stop the execution

```
1   set logging on run-time_traces_log.txt
2   ...
3   ##pre-defined breakpoit##
4   break Initiator_A :: nb_transport_bw
5   commands
6    printf ''instance_name is :''
7    print this ->m_name
8    ...
9    gdb_Initiator_A_nb_transport_bw
10  end
11  ##pre-defined GDB function##
12  def gdb_Initiator_A_nb_transport_bw
13   info line *$pc
14   set $end_func_line_num= $_
15   ...
16   break +1    ##local breakpoint##
17   commands
18    gdb_Initiator_A_nb_transport_bw
19    ...
20   end
21   ...
22   printf ''trans_ID Initiator_A.nb_transport_bw_-_trans_ID is : ''
23   print &trans
24   printf ''trans_phase Initiator_A.nb_transport_bw_-_phase.m_id i
         : ''
25   print phase.m_id
26   ...
27  end
```

Fig. 3.7 A part of the *GDB Command File* of the *LT_AT_BUS*

```
1   Symtab for file 2-stage-pipe.cpp
2   ...
3   Blockvector:
4   ...
5   block #009, object at 0x2f95d20 under 0x2f99948, 3 syms/
        buckets in 0x40ea88..0x40eb1f, function M1::func1()
6    struct M1 : public sc_core::sc_module {
7     sc_core::sc_in<double> in1;
8     sc_core::sc_in<double> in2;
9     sc_core::sc_out<double> out1;
10    sc_core::sc_out<double> out2;
11    ...
12    sc_core::sc_in<bool> clk;
13   public:
14    void func1(void);
15    M1(sc_core::sc_module_name);
16    } * const this; computed at runtime
17    double temp3; computed at runtime
18    ...
```

Fig. 3.8 A part of the debug symbol of the *2-stage pipe* VP generated by GDB

```
1   break func1    ## pre-defined breakpoit ##
2   commands
3     gdb_func1
4   end
5   def gdb_func1    ## pre-defined GDB function ##
6     printf ''temp3 is : ''
7     print temp3
8     break +1        ## local breakpoint ##
9     commands
10      gdb_func1
11    end
12    continue
13  end
```

Fig. 3.9 GDB script to trace variable *temp3* of function *func1*

whenever *func1* is called. That breakpoint is determined automatically via the previous analysis of the debug symbols and called a *pre-defined breakpoint*. The required actions that need to be executed after hitting this breakpoint are defined in a function that is called a *pre-defined GDB function* (lines 5–11). This function is called in the command of the *pre-defined breakpoint* (line 3). To keep tracing the value throughout the *func1* function, additional breakpoints would need to be set for each line of the function (or at least each line that alters variable *temp3*). However, statically setting breakpoints for all potentially interesting lines is not an option. Instead, the breakpoint's commands include an instruction to set up another breakpoint for the next line (line 8) that, again, executes the same commands as previously (lines 9–11). This breakpoint is called *local breakpoint*.

3.4.1.2 Run-Time Information Retrieval

In order to retrieve the run-time information with respect to the execution's flow, the GCF contains a set of breakpoints to pause the execution, store the detailed information of the execution's state, and resume it afterward. As mentioned earlier, these breakpoints are set for each function of the design's modules and the global functions of the model (i.e., *pre-defined breakpoints*). The *pre-defined breakpoints* (which are set up before execution) are placed at any relevant function's first line and thus triggered when their corresponding functions are called. Due to limitations concerning the number of breakpoints, successive lines within these functions cannot all be prepared with breakpoints before the execution as well. Instead, to record any changes within a function, setting a new breakpoint for the next instruction of the function is part of the set of commands that are executed for the first breakpoint, just like it is part of this new set breakpoint. This recursive process is performed repeatedly until the execution reaches the end of the function's body. The goal of a *pre-defined breakpoint* is to halt the execution at the beginning of a module's function, while successive *local breakpoints* are used to step through the body of the module's function line by line. The following examples show how this process is performed to extract the run-time information for each type of VPs, i.e., TLM and cycle-accurate models.

Example 1 In the case of TLM-2.0 VP, consider the *LT_AT_BUS* design. To trace the transactions' behavior of the *nb_transport_bw* function, a *pre-defined breakpoint* is set at the beginning of the function body (Fig. 3.7, line 4). The command of this breakpoint consists of two parts: (1) instruction to extract some structural data (e.g., instance name of module) that cannot be retrieved during static analysis at first phase (Fig. 3.7, lines 6 and 7) and (2) a *pre-defined GDB function* to extract the behavior (e.g., value of transactions' attributes) of the design (Fig. 3.7, lines 12 and 27). After starting the execution of the model, the *pre-defined breakpoint* (Fig. 3.7, line 4) is hit and the corresponding instruction (Fig. 3.7, lines 6 and 7) to extract the instance name of *Initiator_A* module is executed. Afterwards, the *pre-defined GDB function* is called (Fig. 3.7, line 9) to capture the behavioral information, e.g., transaction reference address (Fig. 3.7, lines 22 and 23). It also sets a *local breakpoint* for the next instruction (of function *nb_transport_bw*) to be executed. This process continues until the execution reaches the end of function *nb_transport_bw*.

Example 2 In the case of analyzing cycle-accurate model of SystemC VPs, consider the *func1* function of the *2-stage pipe* VP in Fig. 3.2 (lines 21–33). Here, we show how the run-time value of a local variable within a function can be traced and logged. After starting the execution of the VP under control of GDB, first, the *pre-defined breakpoint* is hit and the execution is stopped at the beginning of function *func1* (Fig. 3.2, line 21). The breakpoint command is executed which is a call to the *gdb_func1* (that is defined to print the value of *temp3* and create local breakpoints). By executing *gdb_func1*, first the value of *temp3* is printed. Then, a break point is set to line 22 of function *func1* (Fig. 3.2). Afterward, instruction *continue* (Fig. 3.9, line 12) in the GDB script is executed. Then, the execution is stopped at line 22 of function *func1* as the *local breakpoint* (which has already set) is hit. Thus, the local breakpoint's script is a recursive call to *gdb_func1* which prints the value of *temp3* and set a new local breakpoint for next line of the function *func1*. This process is repeated until the execution reaches the end of *func1* function (Fig. 3.2, line 33).

3.4.2 Compiler-Based Approach

Figure 3.10 provides an overview of the proposed approach consisting of two main phases as the following.

1. Analyzing the AST of a given VP to extract the static information of the model which is required to describe the design's structure.
2. Generating an instrumented version of the VP's source code using the extracted static information from the previous phase to retrieve the run-time information

Fig. 3.10 The proposed compiler-based information extraction approach overview

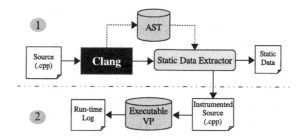

(i.e., behavior). This is performed by automatically compiling the instrumented source code with a standard C++ compiler (e.g., GCC or Clang) and executing it to log the run-time information.

3.4.2.1   Phase 1: Extracting Static Information

The extraction process is performed by visiting relevant nodes in the AST. Clang's AST nodes are modeled on a class hierarchy that does not have a common ancestor. It means that there are multiple larger hierarchies for basic node types such as *Type*, *Decl*, *DeclContext*, or *Stmt*. Many important AST nodes derive from the aforementioned node types with some classes deriving from both *Decl* and *DeclContext*. As the top-level entities of a VP are modules and global functions, the first entry point of extracting design's structure is to find the node including the information of the aforementioned entities.

Thus, once the AST is generated by Clang, it needs to be traversed properly to access the relevant nodes which describe VP's structural information. The first step of this analysis is to identify and access the node types of the AST which are corresponded to the top-level entities of the SystemC VP. The primary node visitor *RecursiveASTVisitor* of Clang provides designers with a recursive mechanism on the entire AST to visit each node based on the *Depth-First Search* (DFS) algorithm. Therefore, it is possible to start from a specific node type related to the top-level entity (e.g., a SystemC module) of the SystemC VP and then recursively traverses the child nodes of the parent node to reach the other constructs (such as signals and ports) which are defined in the top-level entity.

For example, the class reference *CXXRecordDecl* of the Clang is used to represent the C++ *class*, *struct*, or *union*. Therefore, all nodes in the AST that are tokenized by this type represent the aforementioned C++ constructs. As a SystemC module can be defined either using SC_MODULE macro or directly by C++ *struct*, the node type *CXXRecordDecl* is used to access the SystemC module in the AST. In the same way, the node type *FunctionDecl* can be used to find the deceleration nodes related to the VPs' functions (global or local) in the AST.

After extracting the top-level entities, the next step is to retrieve the child elements of them. The information related to the modules' attributes (such as sockets, ports, or signals) and their member functions is also retrieved by visiting the corresponding nodes in the AST. The former can be accessed by visiting the node type *FieldDecl* while the later by *CXXMethodDecl*. In the case of global and local functions, the reference classes *ParamDecl* and *VarDecl* of Clang are used to define function arguments and local variables of functions, respectively. Thus, to access the child elements of functions, it needs to visit the node types *ParamDecl* and *VarDecl* in the AST. Using the motivating examples we show how the structural information can be extracted by analyzing the AST of SystemC-based VPs.

Concerning the TLM constructs, Fig. 3.11 shows a part of the generated AST of the *LT_AT_BUS* VP. The top-level entity of this AST is a node of type *CXXRecord-Decl* representing a SystemC module (lines 1 and 2) which is *Initiator_A*. Analyzing the child element of these nodes shows that the module has only one node of type *FieldDecl*, referring to a simple initiator socket which called *socket* and has 32-bit bandwidth (line 4). It also illustrates that *Initiator_A* is a SystemC TLM-2.0 initiator module as it only has initiator socket.

Regarding analyzing SystemC cycle-accurate models, consider the *2-stage pipe* VP. Figure 3.12 illustrates a part of the generated AST of the VP by Clang. Same

```
1   CXXRecordDecl 0x5c9ce00 <...> line:101:8 struct Initiator_A
         definition
2   |- public 'class sc_core :: sc_module '
3   |- CXXRecordDecl 0x5c9cf50 <col:1, col:8> col:8 implicit
         referenced struct Initiator_A
4   |- FieldDecl 0x5d226e8 <line:104:3, col:43> col:43 referenced
         socket 'tlm_utils :: simple_initiator_socket <Initiator_A
         >':'class tlm_utils :: simple_initiator_socket <struct
         Initiator_A , 32, struct tlm :: tlm_base_protocol_types >'
5   ...
```

Fig. 3.11 A part of the AST of the *LT_AT_BUS* VP generated by Clang

```
1   CXXRecordDecl 0x...20 <...> line:39:8 struct M1 definition
2   |- public 'class sc_core :: sc_module '
3   |- CXXRecordDecl ... col:8 implicit referenced struct M1
4   |- FieldDecl ... col:16 in1 'sc_in <double >':'sc_core :: sc_in <
         double >'
5   |- FieldDecl ... col:16 in2 'sc_in <double >':'sc_core :: sc_in <
         double >'
6   |- FieldDecl ... col:17 out1 'sc_out <double >':'sc_core ::
         sc_out <double >'
7   |- FieldDecl ... col:17 out2 'sc_out <double >':'sc_core ::
         sc_out <double >'
8   ...
9   |- FieldDecl ... clk 'sc_in <_Bool >':'class sc_core :: sc_in <
         _Bool >'
10  |- CXXMethodDecl ... col:10 used func1 'void (void)'
11  ...
```

Fig. 3.12 A part of the AST of the *2-stage pipe* VP generated by Clang

as the previous example, the top-level entity of this AST is also a node of type *CXXRecordDecl*, referring to the SystemC module *M1* (lines 1 and 2). The child nodes of this node present the module's attributes and member functions (lines 4–9). The modules input and output ports are tokenized by node type *FieldDecl* which are *in1*, *in2*, *out1*, *out2*, and *clk*. The member function of the module is *func1* that is identified using node type *CXXMethodDecl*.

The extracted information in this phase is stored in an internal data structure to be used as a foundation of extracting run-time data (i.e., tracing transactions or variables values) and the definition of the design's structure.

3.4.2.2   Phase 2: Extracting Run-Time Information

In order to extract run-time information of a given VP, an instrumented version of the existing source code is automatically generated from the AST including *retrieving* statements. The statements are defined based on a hierarchical structure where for tracing, e.g., a transaction, the value of transaction's attributes and its related parameters such as timing annotation, phase (e.g., BEGIN_REQ), and functions' return status (e.g., TLM_COMPLETED)—for the AT model—are retrieved during execution. Moreover, the simulation time is extracted to notify the exact time of the transaction or variable value changes. To trace a transaction (or variable) after any possible change, we define two locations DEF and USED.

The DEF location refers to the line of code where the transaction (or variable) is defined or declared (e.g., as function arguments or local variables within the function's body). It is extracted by detecting the node type *DeclStmt* and its child node of type *VarDecl* in the AST.

For example, Figs. 3.13 and 3.14 show a part of the AST related to the *LT_AT_BUS* and *2-stage pipe* VPs, respectively. The root node of both AST has node type *CompoundStmt*, referring to the body of functions *thread_process* (Fig. 3.15, line 4) and *func1* of the *Initiator_A* and *M1* modules, respectively. In the first figure (lines 1 and 2), the deceleration node related to variable *status* with type *tlm_sync_enum* is illustrated. As this node is a child element of the *CompoundStmt* node, it is a local variable of the *thread_process* function. Due to its type, the variable is used to store the return status of the TLM interfaces calls. Thus, its value is important to be extracted and traced as it requires to specify the transaction's flow and type. The second figure (lines 2 and 3) shows the deceleration node related to the variable *temp3* with type *double* which is a local variable of function *func1* of module *M1*. Since we want to trace all variables of even local functions for SystemC cycle-accurate model, the value of this variable needs to be extracted.

In the case of TLM-2.0 designs, the USED location refers to function calls (e.g., transport interfaces *b_transport* or *nb_transport*) where the transaction object is used as an input argument. In the case of SystemC designs, USED points to two locations in the source code. First, to access the value of local and global variables which do not require SystemC interfaces to read or write data, the USED location refers to the line of code where the variable is used at the left-hand side

```
1    CompoundStmt 0x4d13bf8
2    ...
3    |-DeclStmt 0x4d0bb68
4    | '-VarDecl ... used status ... 'enum tlm::tlm_sync_enum'
5    |-BinaryOperator ... 'enum tlm::tlm_sync_enum' lvalue '='
6    | |-DeclRefExpr ... 'enum tlm::tlm_sync_enum'... 'status'
         ...
7    | '-CXXMemberCallExpr 0x4d0e818 'enum tlm::tlm_sync_enum'
8    |   |-MemberExpr ... '<bound member function type>' ->
         nb_transport_fw
9    |   ...
10   |   |-UnaryOperator ... 'class tlm::tlm_generic_payload'...
         prefix '*'
11   |   | '-ImplicitCastExpr 0x4..0 'tlm::tlm_generic_payload *'
12   |   |   '-DeclRefExpr ... 'trans' 'tlm::tlm_gen...d *'
13   |   |-DeclRefExpr ... 'class tlm::tlm_phase' ... 'phase'
14   |   '-DeclRefExpr ... 'class sc_core::sc_time' lvalue Var
         ... 'delay'
15   ...
```

Fig. 3.13 A part of the AST of the *LT_AT_BUS* VP related to a non-blocking transport in module *Initiator_A*

```
1    CompoundStmt 0x7f2fbe625448
2    |-DeclStmt 0x7f2fbe624d60
3    | '-VarDecl 0x7f2fbe624d00  used a 'double'
4    | ...
5    |-BinaryOperator ... 'double' lvalue '='
6    | |-DeclRefExpr ... 'double' lvalue Var 0x7...0 'temp3''
         double'
7    | '-ImplicitCastExpr ... 'double' <IntegralToFloating>
8    |   '-ImplicitCastExpr ... 'data_type':'double' <
         LValueToRValue>
9    |     '-CXXMemberCallExpr ... 'const data_type':'const double
         ' lvalue
10   |       '-MemberExpr ... '<bound member function type>' .read
11   |         '-ImplicitCastExpr ... sc_core::sc_in<double>'
         lvalue
12   |           '-MemberExpr ... sc_core::sc_in<double>' lvalue
         ->in2
13   |             '-CXXThisExpr ... 'struct M1 *' this
14   ...
```

Fig. 3.14 A part of the AST of the *2-stage pipe* VP related to line 30 of Fig. 3.1

of an assignment statement (e.g., expression or function call). Second, regarding the module ports (e.g., signals with type *sc_in* or *sc_out*) as they use the *read()* or *write()* interfaces to access or modify data, the USED location points line of code where a module port reads or writes data. Thus, the *retrieving* statements are inserted to the source code after the aforementioned locations. The following examples show how

```
1    struct Initiator_A : sc_module{
2    tlm_utils :: simple_initiator_socket <Initiator_A , 32> socket;
3    ...
4    void thread_process (){
5    tlm :: tlm_generic_payload* trans ;
6    tlm :: tlm_phase phase ;
7    sc_time delay ;
8    ...
9    status = socket->nb_transport_fw (*trans , phase , delay );
10   Fout<< "Initiator_A :: thread_process :: trans . ID="<< trans << "
         DATA="<< trans -> get_data_ptr ()<< "CMD="<< trans ->
         get_command ()<< "ADR="<< trans -> get_address ()<< "RSP="<<
         trans -> get_response_status ()<< "DL="<< trans ->
         get_data_length ()<< "delay="<< delay << "phase="<< phase << "
         instance_name_module="<< this ->name ()<< "ST="<<
         sc_time_stamp ()<< endl ;
11   ...}
```

Fig. 3.15 A part of the instrumented source code of module *Initiator_A* of the *LT_AT_BUS* VP

this process is performed and the aforementioned locations are extracted from the AST representation of a given SystemC-based VP.

Example 3 Consider a part of the source code related to the module *Initiator_A* of the *LT_AT_BUS* VP (Fig. 3.15). Line 10 is not initially available. Assume that we want to trace all transactions generated by the *thread_process* function of the *Initiator_A* module. To do this, the VP's AST is analyzed by *Static Data Extractor* module (Fig. 3.10, phase 1) to find DEF and USED locations in the source code. For example, consider a USED location (line 9, in Fig. 3.15) where transaction *trans* is used as a function argument of the *nb_transport_fw* interface. To properly trace the transactions, all information related to the transactions' flow, data, and type must be extracted. This includes

1. the module name (*Initiator_A*) and the parent function (*thread_process*) to which this transaction belongs and the transactions' reference address,
2. all attributes of the transactions which are data, address, response status, data length, and
3. transactions' related parameters which are the *phase* and *delay* arguments of the *nb_transport_fw* interface and its return status stored in the *status* variable.

As illustrated in Fig. 3.13 (line 7), the corresponding node in the AST of the VP for this USED location is the node of type *CXXMemberCallExpr*. It shows that the communication interface *nb_transport_fw* is called that has the return type *tlm_sync_enum* and three input arguments *trans* (line 12), *phase* (line 13), and *delay* (line 14) with types *tlm_generic_payload*, *tlm_phase*, and *sc_time*, respectively. The input arguments of the *nb_transport_fw* are the child elements of the *CXXMemberCallExpr* node that has the node type *DeclRefExpr*. The other important element that needs to be extracted for this USED location to properly

generate the *retrieving* statement is to extract the return value of the *nb_transport_fw*
interface. As the *CXXMemberCallExpr* node is the right-hand side child element of
the *BinaryOperator* "=" (line 5), we need to find the left-hand side child element of
this node with type *DeclRefExpr* referring to a variable with type *tlm_sync_enum*
(line 6).

From the extracted information, the *retrieving* statement *Fout* (line 10, Fig. 3.15)
is automatically generated and inserted after the USED location in the new source
code. The instructions *this->name()* and *sc_time_stamp()* are also added to the
retrieving statement to identify that the transaction *trans* belongs to which instance
of the *Initiator_A* module and the simulation time when the transaction is sent
through the initiator socket *socket*, respectively.

Example 4 Consider the module *M1* of the *2-stage pipe* VP (Fig. 3.2). Assume that
we want to trace the local variable *temp3* of the *func1* function. To do this, the
AST of design is analyzed by *Static Data Extractor* module to find DEF and USED
locations in the source code. For example, consider a USED location (line 30,
in Fig. 3.2) where variable *temp3* is used in the left-hand side of an assignment
expression. The left-hand side expression of this node is variable *temp3*. The parent
function (*func1*) and module (*M1*) to which this variable belongs are also retrieved.

The aforementioned location falls into the first category of the USED location
for the SystemC cycle-accurate model where the goal is to find the assignment
statements. This is performed by finding the node of type *BinaryOperator* indicating
an assignment operator "=" (Fig. 3.14, line 5). The next step is to find the left-hand
side child node of type *DeclRefExpr* referring to the local variable *temp3* (Fig. 3.14,
line 6). From the extracted information, the *retrieving* statement *Fout* (Fig. 3.16,
line 5) is automatically generated and inserted after the USED location in the new
source code. The instructions *this->name()* and *sc_time_stamp()* are added to the
retrieving statement to identify that the variable belongs to which instance of module
M1 and the simulation time when a new value is assigned to a variable, respectively.

For both aforementioned examples, the *Rewriter* interface of Clang is used to
insert the *retrieving* statements into the corresponding design's lines of code (DEF
or USED locations) and generate its new instrumented version.

In summary, the most important Clang constructs that are used for our analysis
in this section are presented in Table 3.1. These constructs were used to build the
Static Data Extractor module of the proposed approach. Table 3.1 also provides

```
1    void M1::func1(){
2      double    temp1, temp2, temp3;
3      ...
4      temp3 = genA + in2.read()*genB;
5      Fout<<"M1::func1::temp3 = "<<temp3<<" instance_name_module:
           "<<this->name()<<" simulation_time: "<< sc_time_stamp()
           <<endl;
6      ...}
```

Fig. 3.16 A part of the instrumented source code of function *func1* of the *2-stage pipe* VP

Table 3.1 The relationship between different node types in the AST and the corresponding SystemC constructs for a given VP

AST node type	SystemC constructs
CXXMethodDecl	SystemC process
FieldDecl	SystemC ports and sockets
CXXMemberCallExpr	SystemC interface (e.g., nb_transport)
DeclRefExpr	Local and global variable
FuncDecl	Local and global function
CXXRecordDecl	SystemC module

a connection between different node types in the AST of a given VP and the corresponding SystemC constructs that need to be extracted for the proposed SystemC VP analysis approach.

3.4.3 Designer Interaction for Optimized Information Extraction

In order to empower designers to interactively customize the proposed approaches for their specific use cases or applications (e.g., design understanding or verification tasks), the proposed approaches contain a set of configuration options and parameters in the first phase of their analysis. By this, the *GDB Command File* (GCF script) or a given VP's instrumented source code can be generated based on the designers' optimization. These interactive options are as follows:

- filtering some elements of the design, e.g., only trace certain modules (and their signals),
- filtering based on the depth of information, e.g., trace only modules (and their member functions and variables),
- filtering some specific types of function, variables or signals, e.g., do not trace global functions, local variables, or signals of type *sc_in*.

This allows designers to filter out the information that is not required for their analysis and focus on the points of interest.

3.5 Visualization

After extracting the required information from a given SystemC-based VP, the next step is to present this information in such a way that designers can quickly grasp the different aspects of the VP's structure and behavior. This post-execution step is considered as visualization which is an important issue for several tasks during the design process. It can be used to help, e.g., designers to understand the complexity of a system, to locate errors in running systems, or to illustrate the project's documentation. Therefore, the main difficulty that needs to be overcome in this section is how the extracted information should be visualized and presented that empower designers to easily

- trace both structure and behavior of the VP and
- use it for tasks of debugging, validation, and verification off-the-shelf or with minimum translation effort into their desired forms.

Conventional approaches (discussed in Chap. 2) mostly focus on presenting the extracted information in an IR model such as XML or AST. Although it can be a good solution to present the structure of a VP, to describe its behavior, the results may become very complex to be understood especially when the complexity of VPs increases. In the RT level, the common solution to monitor the behavior of a cycle-accurate model (implemented using Verilog or VHDL) is the utilization of the *Value Change Dump* (VCD) file. This provides designers with a waveform that trace, e.g., the state of a single variable during the simulation time of design. This option is also provided by the standard SystemC API (i.e., using *sc_trace*). Although this method works well for SystemC data types which are defined as signals of modules, it comes with several drawbacks as the following.

- It lacks precision for base type variables (e.g., C++ data type) that may change several times during a single SystemC-δ-cycle (the smallest amount of time that may pass concerning the simulation kernel).
- It fails for user-defined datatypes that are not supported at all unless the designers alter their code.
- It cannot trace the values of local variables of modules and functions.
- It requires further programming effort by designers to manually modify the source code and include all signals that need to be trace.

Therefore, to present the behavior of cycle-accurate SystemC VPs, the aforementioned drawbacks must be overcome for generating VCD files.

In the case of SystemC TLM-2.0 VPs, a simple waveform does not provide designers with a comprehensive presentation of the VPs' behavior. The main reason is that a transaction is a complex type (not a simple variable) and it is very hard to describe the extracted information related to its flow, data, and type in the shape of a waveform. The other important challenge related to the behavioral understanding of the TLM-2.0 models is the communication interface. The VP models communicate through thousands of TLM-2.0 transactions which are difficult to follow, redundant, and unnecessary from the perspective of design understanding. A concise visualization of the underlying VP transactions flow can accelerate the revisions and additions while helping significantly in design understanding. The TLM-2.0 reference manual in [5] suggests UML message sequence chart for the transactions' behavior presentation. However, it only shows the transaction flow and not its data and type. Hence, we present the behavior of VPs in terms of UML diagram inspired by [5] that in addition to the transaction's flow, its data and type are covered as well.

3.5.1 VP Structure Presentation

The preferred format to present the VPs' structure depends on the purpose of designers or the back-end tools which may use the results for further analysis. Because we want to have a generic presentation of this information, we have chosen to store all information extracted from the model in an XML document. The XML format can be easily read and parsed by future tools which may use the result of our analysis, while at the same time provides designers with a structural presentation of the design architecture.

The structure of a given VP model is extracted in two steps: (1) during the static analysis in the first phase of the proposed approaches (Figs. 3.5 and 3.10) including

- the root name and type of each module (due to the TLM-2.0 reference manual for TLM modules, a module's type can be initiator, interconnect, or target. This is identified by analyzing its sockets type),
- the name and type of each function,
- the variables of each module, and
- local variables of each function

and (2) in the second phase of the proposed approaches (Figs. 3.5 and 3.10) during the execution of the VP including

- the instance name of each module and
- binding information of signals and sockets.

The extracted information from both steps is bound together to create a complete structure of the design. As shown in Fig. 3.17, this is performed by the *Static Data analyzer* module. The root element of the generated XML document is the name of the VP model. The structure of the VP is hierarchical itself, with the first child elements being modules and global functions. The child elements of these are their respective member functions and attributes. For example, Fig. 3.18 illustrates a part of the generated XML document of *LT_AT_BUS* VP. The root element of this XML document is the name of VP (line 1). The structure of the VP is identified by a set of child elements of the root element which are modules or global functions (line 3). The child elements of each module or global function are their respective member functions and attributes (lines 4–14).

Fig. 3.17 Overview of the VPs' structure presentation approach

```
1   <TLM_Design_Architecture Design_name = "LT_AT_BUS">
2    ...
3    <Module_name name = "Target_A" type = "target">
4     <Function name = "send_response" type="void">
5      <Local_var name = "trans" type = "tlm_generic_payload"/>
6      <Local_var name = "status" type = "tlm::tlm_sync_enum"/>
7      <Local_var name = "bw_phase" type = "tlm_phase"/>
8      <Local_var name = "delay" type = "sc_time"/>
9     </Function_name>
10    ...
11    <Global_var name = "response_in_progress" type = "bool"/>
12    <Global_var name = "n_trans" type = "int"/>
13    <Global_var name = "socket" \
14    type="tlm_utils::simple_target_socket"/>
15    ...
16   </TLM_Design_Architecture>
```

Fig. 3.18 A part of the structural presentation of the *LT_AT_BUS* VP

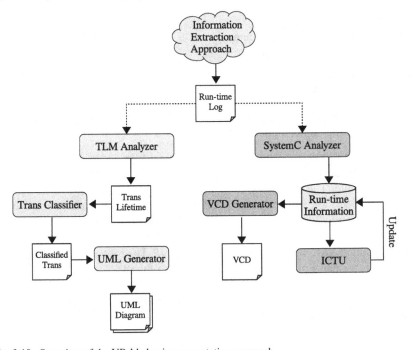

Fig. 3.19 Overview of the VPs' behavior presentation approach

3.5.2 VP Behavior Presentation

The generated *run-time log* file (Fig. 3.19) by the information extraction approaches (either debugger-based or compiler-based) contains unordered information about the VPs' behavior. As the information is stored in the order of execution, transactions (or value of signals and variables) overlap in this log file as method or function calls related to a particular task may be carried out at different points in time. To present the behavior of each particular transaction (or value of a variable), this large set of data must be separated into sets that each refers to a single given transaction (or variable).

3.5.2.1 Behavior Presentation for SystemC Designs

One of the common solutions to trace the simulation behavior of cycle-accurate designs (e.g., systems modeled at the RTL using Verilog or VHDL) is the utilization of VCD files. To present the value of SystemC signals, the standard SystemC API uses the VCD format for the cycle-accurate models. The SystemC API uses SystemC δ cycle notion as the smallest timestep to differentiate changes on each signal in the VCD file. This may result in less precision for base type variables. As in the proposed approaches we take advantage of GDB as an execution environment or Clang to modify the source code, the precision of information extraction is C++ statements (not only SystemC primitives). Thus, the smallest timestep to differentiate assignments is a C++ statement. This (potentially) increases the precision of tracking value changes in the order they occur in, while still setting them in the context of the current simulation time. We call this behavior of values within a single δ cycle the *intra-cycle behavior.*

The *intra-cycle behavior* analysis can facilitate the design process such as debugging where a C++ data type is altered by some wrong computations and then used in the right-hand side to be assigned to a SystemC signal. While the conventional VCD file only illustrates the final value of the signal at each δ cycle, the *intra-cycle behavior* analysis enables designers to trace all local variables' intermediate changes within the δ cycle. Thus, this can help designers to detect the exact point of the bug.

As illustrated in Fig. 3.19, first, the retrieved information is stored in an internal data structure by the *SystemC Analyzer* module where each simulation time stamp is considered as a time unit in which the state of variables that was extracted at this time is stored. Second, to differentiate values of variables within a single point in simulation time, the *Run-time Information* is processed by the *Intra-Cycle Time Unit (ICTU)* module.

To present all events on a single timeline, the assignments of every single point in simulation time are sorted by their execution order o and the value $o \cdot \mu_t$ is added to their original time stamps. The value μ_t is thus used to differentiate the particular assignments. It should be much smaller than the smallest step in simulation time to have all assignments being displayed before the next "large" simulation timestep. The parameter μ_t is therefore related to the maximum sum of the number of value changes of variables in a time unit among all time units as illustrated in Algorithm 3.1 and is calculated automatically.

Algorithm 3.1: ICTU update process

Input: simulation time scale stc, Run-time Information RI
Output: updated RI
1 **foreach** *time unit t in RI* **do**
2 | **foreach** *variable v in t* **do**
3 | | $vc[t][v] \leftarrow sum(uinq\ value\ changes\ v)$;
4 | **end**
5 | $tvc[t] \leftarrow sum(vc[t])$;
6 **end**
7 $max_{tvc} \leftarrow max(tvc)$;
8 $\mu_t \leftarrow (1/max_{tvc}) * stc$;
9 **foreach** *time unit t in RI* **do**
10 | **foreach** *variable v in t* **do**
11 | | **if** $vc[t][v] > 1$ **then**
12 | | | **foreach** *value assignment* **do**
13 | | | | $t_{new} \leftarrow t + \mu_t$;
14 | | | | $store\ (t_{new}, value\ assignment)$;
15 | | | | $t \leftarrow t_{new}$;
16 | | | **end**
17 | | **end**
18 | **end**
19 **end**
20 update (RI);

Finally, the updated version of the *Run-time Information* is used to generate a VCD file of the design's behavior via the *Generate VCD* module. The generated file includes:

- the name of modules, their instances and the global functions in form of the *Signal Search Tree* (SST),
- the state of the design's variables after the simulation, and
- the value of each variable which is assigned during run-time in the shape of a waveform w.r.t the simulation time.

3.5.2.2 Behavior Presentation for TLM Designs

Since the extracted information of each transaction is scattered over the *Run-time Log* file, an information analysis approach is required to reduce the complexity of understanding the extracted information.

The first step of this analysis is to describe each extracted transaction based on its flow, data, and type within its lifetime. This requires an isolation of every single transaction and its corresponding information from other transactions. In order to trace a single transaction in the *Run-time log* file, transactions are separated based on some unique elements. The key element for this isolation is the transaction reference address. We take advantage of the TLM-2.0 rule stated in [5]—a transaction object is passed as a function argument to a method implementing one of the given communication interfaces (*b-transport* or *nb-transport*) with a reference address (call by reference). The reference address of a transaction object remains constant from its creation until its destruction (i.e., during its lifetime). For transactions that are generated by different TLM initiator modules, the reference address of transactions can be used to isolate information related to each of them. Thus, this reference address is used as a *transaction ID*, which is the main key to trace a transaction (related information to build its lifetime) among other transactions within the simulation log of a TLM design.

In addition to the transaction reference address (considered as *transaction ID*), some attributes of a transaction object (e.g., response status) as well as other elements related to it (e.g., the value of the phase argument on call to and return from the *nb-transport* function and the return value of the function) are used to determine the start and endpoint of the transaction. The phase argument represents the current state of a module w.r.t. the TLM-2.0 base protocol state machine of phase transition. This information is referred to as the *transaction-related information*.

As demonstrated in Fig. 3.19, the *TLM Analyzer* module gets the *Run-time log* as an input. It extracts the required information for each single transaction to describe the activity within its lifetime and stores it in the *Trans lifetime* file. This data is an accurate trace of each transaction's behavior, covering all changes in transaction data that occurred during the execution of the model. Figure 3.20 shows a part of the generated *Trans lifetime* of the *LT_AT_BUS* design.

Since it is possible that many of the extracted transactions in *Trans lifetime* have the same flow and type (only their data is different), a further analysis step is required to visualize only those which present a unique behavior. This effectively reduces the number of generated UML diagrams, allowing designers to quickly understand the behavior of a given TLM-2.0 design.

```
1    transID_number  :0x7054f0_1
2    ...
3    ([Target_A::nb_transport_fw, trgA\
4    ,target_a.cpp:74, 0 ns, 0x7054f0\
5    ,target], [0x6bc660, 100, tlm::TLM_READ_COMMAND\
6    ,4, tlm::TLM_OK_RESPONSE, BEGIN_REQ, 4565 ps\
7    ,NULL], [])
8    ...
9    transID_number  :0x7078f0_1
10   ...
11   ([LT_AT_BUS::nb_transport_fw, bus_0\
12   ,LT_AT_BUS.cpp:196, 6397 ns, 0x7078f0\
13   ,interconnect], [0x6c9dc4, 232, tlm::TLM_WRITE_COMMAND\
14   ,4, tlm::TLM_OK_RESPONSE, END_RESP, 243 ps\
15   ,tlm::TLM_COMPLETED], [])
16   ...
```

Fig. 3.20 A part of the *Trans Lifetime* of the *LT_AT_BUS* VP

Classifying Transactions

The transactions' classification is performed in two levels as the following.

- First, based on the transactions' flow providing designers with an abstract view. Essentially, it distinguishes transactions based on different communications pattern related to the TLM modules taking part in their lifetime.
- Second, based on transactions' type providing designers with an accurate analysis of the transactions' type that different TLM modules used to communicate.

In order to classify the transactions based on their flow, the *Trans Lifetime* file is analyzed by the *Trans Classifier* module. This analysis is performed by generating a *Communication Pattern String* (CPS) for each transaction's lifetime stored in the *Trans lifetime*. For a given transaction, the CPS is a string of characters generated by concatenating the root and instance names of all modules taking part in the transaction's lifetime. For example, the CPS for the transactions generated by the *Initiator_A* (Fig. 3.1) to access *Memory_A* through *LT_AT_BUS* is "*Initiator_A:init_0+LT_AT_BUS:bus_0+Memory_A:trgA*." By this, transactions in the *Trans lifetime* are categorized into several sub-groups considered as *Unique Flow Group* (UFG) where each group presents a unique flow (communication pattern).

The next step of the transactions' classification is to perform a transaction type analysis in each UFG. For each transaction of a UFG, first, the type of timing model is identified (LT or AT) based on the type of communication interface. By identifying the types of communication interface (blocking or non-blocking) in each transaction lifetime, the type of timing model is extracted which is either LT, AT, or a combination of both. While the LT model can only be implemented in one way due to the TLM-2.0 base protocol, the AT model requires further analysis as it can be implemented in 13 unique ways (see Table 2.1).

In the case of the AT model, we take advantage of the transactions' related parameters to distinguish different types of base protocol transactions in each group. To do this, a *Transaction Type String* (TTS) is generated for each AT model transaction's lifetime in a unique communication group. The TTS includes a concatenation of the communication interface(s), return status(es), and transition phase(s) of an AT model transaction. For example, the TTS of the T_2 transaction type (Table 2.1) generated by the *Initiator_A* module (Fig. 3.1) is "*fw+TU/TC+BRQ/BRP/ERP*." Therefore, each UFG is divided into several sub-groups wherein each group the transactions have the same TTS. The final classified results are stored in the *Classified Trans* by the *Trans Classifier* module.

Finally, to reduce the complexity of understanding the extracted information stored in the *Classified Trans*, the UML diagram is generated by the *Generate UML* module for a transaction in each UFG that has a unique type. The generated UML diagram is a message sequence chart introduced by the OSCI TLM-2.0 reference manual in [5] but it provides more detailed information. It describes the transaction flow among modules' communication for each single transaction within the design. It also includes the transaction data (i.e., the last changes of the transaction data

Table 3.2 The transaction classification results of the *LT_AT_BUS* VP

Number	Flow	Modules	Transaction type
1	F_1	Initiator_A→LT_AT_BUS→Memory_A	T_0
2	F_1	Initiator_A→LT_AT_BUS→Memory_A	T_4
3	F_2	Initiator_A→LT_AT_BUS→Memory_B	T_1
4	F_3	Initiator_A→LT_AT_BUS→Memory_C	T_2
5	F_4	Initiator_B→LT_AT_BUS→Memory_A	T_3
6	F_5	Initiator_B→LT_AT_BUS→Memory_B	T_3
7	F_6	Initiator_B→LT_AT_BUS→Memory_C	T_3

and not all temporary changes), which is passed among modules during their interactions. In particular, the UML model includes a set of sequence numbers indicating both transaction flow and transaction data within its lifetime.

For example, Table. 3.2 shows the transaction classification results related to the *LT_AT_BUS* VP. It illustrates that the VP has six different communication flows, four different transaction types, and overall seven unique patterns of flow and type. Thus, seven UML diagrams are generated to present the simulation behavior of the VP. Figure 3.21 illustrates the UML diagram of an extracted transaction of type T_0 generated by the initiator module *Initiator_A*. The black shapes present root and instance name of modules within the design. The role (type) of each module is shown on top of the modules' name. The information on each arrow demonstrates the interaction between two modules that are drawn from the caller to the callee w.r.t the simulation time. In particular, for a call from an instance of a TLM module, it presents the number of the sequence, the name of the caller function, and timing annotation. Moreover, the generated UML model includes detailed transaction data. The box under each arrow shows the transaction's attributes which are passed as an argument from the caller to the callee. The white boxes illustrate a local transaction object, while the blue boxes demonstrate a transaction object reached the callee through a function call. For example, *seq-2* in Fig. 3.21 contains the information related to the response of the target module *Memory_A*. This information is the name of the called function (*exec_func*) and timing annotation (*20 ns*). It also includes the transaction data passed to the callee module (*LT_AT_BUS*) including the reference address of the transaction (*0x7bb3c3*), address (*Ox09*), command (*tlm::TLM_READ*), length (*4*), and response status (*tlm::TLM_OK_RESPONSE*).

3.5.3 Designer Interaction for Optimized Translation

In order to provide a better view of a given ESL design's behavior and structure, the current implementation offers several configuration options available for designers. The extracted information of the design can be filtered based on the following parameters:

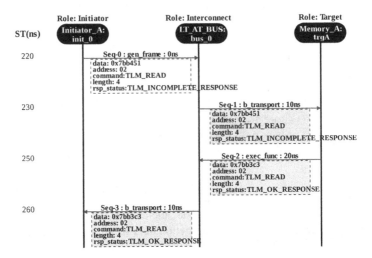

Fig. 3.21 UML diagram related to a transaction of type T_0 of the *LT_AT_BUS* VP

- kinds of variables: the generated VCD file is optimized only to show the signals of modules, local variables of modules' functions, or variables of global functions,
- depth of information: the generated VCD file may exclude the *intra-cycle behavior* (resulting in a more "classic" SystemC trace that only tracks values when the simulation time advances),
- time window: the generated VCD shows only the information from a specific period of simulation time, and
- depth of hierarchy: the generated XML model is optimized only to show the information of modules (and also their member functions and their variables).

In the case of SystemC TLM-2.0 designs, the generated UML model may only show the sequence of specific modules' activity or include the transactions' data as well. This feature allows designers to only focus on the points of interest in the design by filtering out any irrelevant information. Moreover, reducing the amount of translated information enhances the readability of generated UML, VCD, or XML files.

3.6   Experimental Evaluation

This section presents the experimental results obtained using both of the suggested approaches in this chapter. Several standard ESL VPs from various domains have been used to evaluate the quality of the proposed approaches. The experimental evaluation is presented in two steps as the following. First, the quality of the proposed approaches to extract the required information from both of the SystemC cycle-accurate and TLM-2.0 VP models is illustrated in Sect. 3.6.1. Moreover, to demonstrate how the structure and the behavior of SystemC VPs are visualized based on the extracted information, two case studies which are *RISC-CPU* (implemented in SystemC cycle-accurate model) [1] and *AT-example* (implemented in SystemC TLM-2.0) [6] are discussed in details. Second, we give a brief discussion based on the obtained results to evaluate the characteristics, performance, and memory usage of the proposed approaches in Sect. 3.6.2.

The experiments are carried out on a PC equipped with 8 GB RAM and an Intel Core i7 CPU running at 2.4 GHz.

3.6.1 Case Studies

This section shows the experimental results of applying the proposed approaches for each type of SystemC VPs. The detailed analysis of SystemC cycle-accurate and TLM-2.0 VP models is presented in Sects. 3.6.1.1 and 3.6.1.2, respectively.

3.6.1.1 SystemC Cycle-Accurate VPs

The SystemC VPs presented in Table 3.3 are taken from the standard designs which provided by OSCI [1], S2CBench [97], GitHub [30, 86], and the University of Edinburgh [54]. In this table, column *Test Size* illustrates the size of the application running on each design. Columns *SystemC Model, LoC, Comp,* and *Test* show the name of designs, the lines of code, the number of modules, and the number of tests applying to each design, respectively. We compared the results of our proposed approaches to the SystemC trace API in terms of two important parameters which are required to analysis SystemC cycle-accurate VP models. These parameters are the number of variables *#Variable* that are traced during execution time and the number of extracted time units *#TimeUnit*.

As demonstrated in Table 3.3, the amount of both traced variables and time units for all case studies of the proposed approaches are the same or much higher than those traced via the SystemC trace API. The parameter *#TimeUnit* represents value changes of variables. These parameters illustrate the accuracy of our approaches to extract the detailed behavior of a SystemC design. This difference in the amount of traced data reflects one significant advantage of the proposed approaches as it enables designers to trace value changes even for both, global and local native variables (that have the SystemC primitive data types) or compound data types (which can be constructed using the programming language's primitive data types). The proposed approaches can trace variables, regardless of whether they are placed on the stack or the heap not only when the simulation time advances but also of any assignments in-between. Moreover, using the SystemC trace API to analyze the behavior of a given SystemC model is an intrusive solution. It required further programming effort by designers to modify the source code to include all variables

Table 3.3 Experimental results related to the amount of extracted information for all SystemC cycle-accurate VP models

Test size	SystemC model	LoC	#Comp	#Test	SystemC trace API		Compiler-based approach		Debugger-based approach	
					#Variable	#TimeUnit	#Variable	#TimeUnit	#Variable	#TimeUnit
Small	4-bit shift-register[a]	135	2	50	3	29	12	54	12	54
	FIR-filter[b]	233	5	30	6	24	23	504	23	504
	3-stage-pipe[b]	290	5	10	8	23	25	34	25	34
	FFT_flpt[b]	484	3	10	8	81	34	129	34	129
	Cholesky[c]	522	2	27	4	11	28	46	28	46
	Hamming-code[d]	563	6	16	13	32	55	64	55	64
	Interpolation[c]	596	2	10	10	21	32	30	32	30
	IDCT[c]	815	2	100	9	205	27	206	27	206
	VGA[c]	856	6	100	9	59	37	239	37	239
	Decimation[c]	883	2	20	10	47	35	48	35	48
	pkt-switch[b]	1020	10	50	14	93	332	5606	332	5606
	MD5-hash[c]	1111	2	5	18	23	33	115	33	115
	RISC CPU[b]	1960	10	10	89	37	299	121	299	121
	Simple-bus[b]	2100	7	10	10	359	32	852	32	852
	AES128lowarea[c]	3280	13	5	10	18	86	89	86	89
	LZW-encoder[e]	5132	22	5	20	30	398	139	398	139

Large									
4-bit shift-register[a]	135	2	150	3	78	12	142	12	142
FIR-filter[b]	233	5	150	6	66	23	2291	23	2291
3-stage-pipe[b]	290	5	150	8	300	25	341	25	341
FFT_flpt[b]	484	3	150	8	1465	34	5841	34	5841
Cholesky[c]	522	2	108	4	29	28	429	28	429
Hamming-code[d]	563	6	128	13	256	55	512	55	512
Interpolation[c]	596	2	104	10	205	32	295	32	295
IDCT[c]	815	2	996	9	1997	27	1998	27	1998
VGA[c]	856	6	1000	9	619	37	2005	37	2005
Decimation[c]	883	2	215	10	437	35	438	35	438
pkt-switch[b]	1020	10	350	55	1723	332	19,559	332	19,559
MD5-hash[d]	1111	2	50	18	66	33	307	33	307
RISC CPU[b]	1960	10	150	89	137	299	368	299	368
Simple-bus[b]	2100	7	100	10	4506	32	9767	32	9767
AES128lowarea[c]	3280	13	50	10	110	86	397	86	397
LZW-encoder[e]	5132	22	50	20	290	398	529	398	529

LoC line of code, *#Comp* number of components
[a]Provided by University of Edinburgh [54]
[b]Provided by OSCI [1]
[c]Provided by Schafer and Mahapatra [97]
[d]Provided by GitHub [30]
[e]Provided by orahyn [86]

that need to be traced. Thus, for an intricate design with lots of variables, it might be a time-consuming task.

In the following, as a representative of analyzing a SystemC cycle-accurate model, the result of analyzing the *RISC-CPU* VP obtained by both proposed approaches is presented in detail. This includes the visualization of the VP's structure and run-time behavior with the precision of the intra-cycle behavior.

RISC-CPU

The *RISC-CPU* design is a standard OSCI example implementing a CPU in SystemC using ten different modules. The instruction set is based on commercial RISC processors together with MMX-like instruction for DSP programs. It consists of more than 39 instructions, such as arithmetic, logical, branch, floating point, and SIMD (MMX-like). In order to show the intra-cycle behavior, we modified the *exec* module by adding a combinational function calculating the *factorial* of the *dina* input. The factorial function is added as an exemplary hardware accelerator, computing the factorial computation in a single clock cycle. Integrating such accelerators is a common approach to gain performance for a specific task. Figure 3.22 illustrates a part of the generated VCD file of the *RISC-CPU* system. The retrieved structural and behavioral information of the design is shown in four parts in this figure.

1. It shows a basic hierarchy of the *RISC-CPU* design in the form of the *Signal Search Tree* (SST). It includes the name of modules and their instances as well as the global functions (e.g., *scmain* function in this example). E.g., the *exec* module has only one instance in the *RISC-CPU* design, which is *EXEC_BLOCK*.
2. It illustrates the state of the design's variables after the simulation. Each variable is identified based on its hierarchical structure which consists of the name of the

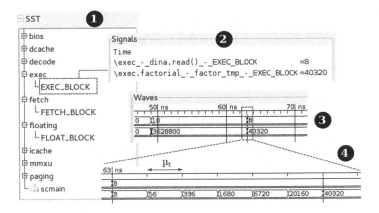

Fig. 3.22 A part of the generated VCD file of the RISC-CPU VP

root module, name of the function (for local variables), name of the variable, and the instance name of the root module. The expression *exec_-_dina.read()_-_EX-EC_BLOCK* shows that the instance *EXEC_BLOCK* of module *exec* has a signal *dina*. The value of this signal (here is eight) is accessed by member function *read()* of its type. The expression *exec.factorial_-_factor_tmp_-_EXEC_BLOCK* illustrates that the function *factorial* of instance *EXEC_BLOCK* of module *exec* has a local variable *factor_tmp*. The value of this signal is 40,320.

3. It shows the value of each variable in the shape of a waveform with respect to the simulation time.
4. It illustrates the intra-cyclic information, showing how new, small timesteps illustrate a variable being increased within a single SystemC-δ-cycle. In this example, at time unit $t = 63$ ns the input signal *dina* is assigned to eight, thus the combinational function *factorial* is called to calculate 8!. It is computed by a *for-loop* statement where the local variable *factor_tmp* is defined to store the partial values of the factorial of *dina* input in each iteration. As the entire calculation is performed in a single SystemC-δ-cycle, these temporary changes are covered by the intra-cycle behavior analysis. The *Dynamic Information* is analyzed by ICTU module based on Algorithm 3.1 to cover all temporary changes in a single time unit. As the maximum amount of assignments within each of the simulation timesteps is $max = 100$ and the simulation time scale is $ts = 1$ ns, the smallest step within a single time unit is $\mu_t = \frac{ts}{max} = 10$ ps. This allows the proposed approaches to store all value changes (even temporary ones).

3.6.1.2 SystemC TLM-2.0 VPs

The SystemC TLM-2.0 VPs in Table 3.4 are the standard design provided by Doulos [6] and [101]. Columns *Test size* and *TLM Model* show the size of the application running on each VP and name of each SystemC TLM-2.0 design, respectively. The *LoC*, *#Comp*, and *#TM* illustrate the complexity and difference of each design in terms of lines of code, number of components, and the timing model, respectively. Column *#Trans* presents the number of transactions that are extracted in each design. For both proposed approaches, columns *#UFlow*, *#UType*, and *#UML* show the number of unique flow, transactions' type (*T*), and generated UML diagrams (*UML*), respectively. As illustrated in the table, both debugger-based and compiler-based approaches provide designers with the same number of extracted transactions' flow and type. Moreover, the visualization phase for both approaches generates the same number of UML diagrams. It means that the proposed approaches have the same accuracy in tracing the detailed behavior of TLM VPs and reducing the complexity of understanding this behavior by generating the same number of UML diagrams. The other important point that needs to be taken into consideration is that the number of generated UML diagrams using the suggested approaches for both small and large number of transactions is the same. It means that they are able to filter out the redundant transactions and only keep those

Table 3.4 Experimental results related to the amount of extracted information for all SystemC TLM-2.0 VP models

Test size	TLM model[a]	LoC	#Comps	TM	#Trans	Compiler-based approach			Debugger-based approach		
						#UFlow	#UType	#UML	#UFlow	#UType	#UML
Small	LT-example[1]	175	2	LT	16	1	1	1	1	1	1
	Routing-model[1]	456	6	LT	10	4	1	4	4	1	4
	Example-4[1]	547	2	AT	10	1	4	4	1	4	4
	Example-5[1]	650	7	LT	10	7	1	7	7	1	7
	Example-6[1]	713	9	AT	20	16	2	16	16	2	16
	AT-example[1]	3410	19	AT	20	12	9	14	12	9	14
	Locking-two[1]	4690	23	LT/AT	20	14	10	16	14	10	16
	SoCRocket[2]	50,000	20	LT/AT	100	19	8	21	19	8	21
Large	LT-example[1]	175	2	LT	160	1	1	1	1	1	1
	Routing-model[1]	456	6	LT	100	4	1	4	4	1	4
	Example-4[1]	547	2	AT	348	1	4	4	1	4	4
	Example-5[1]	650	7	LT	69	7	1	7	7	1	7
	Example-6[1]	713	9	AT	245	16	2	16	16	2	16
	AT-example[1]	3410	19	A T	49	12	9	14	12	9	14
	Locking-two[1]	4690	23	L T/AT	371	14	10	16	14	10	16
	SoCRocket[2]	50,000	20	L T/AT	1000	19	8	21	19	8	21

LoC line of code, *TM* timing model, *#Trans*, number of transaction, *#UFlow* number of unique transaction flow, *#UType* number of unique transaction type, *#UML* number of generated UML diagram

[a]Please note that the VP models are modified to support more variety of transactions type and flow than the original models

[1]Provided by Aynsley [6]

[2]Provided by Schuster et al. [101]

which are required to present the unique behavior of a given VP. This can effectively reduce the effort of designers to understand the VP's complexity. However, to accurately visualize the complete behavior of the VP (different communication patterns and transaction types), the running application (benchmark or software) must properly activate different module of the VP to cover all transaction's flow and type. For example, the upper part of Table 3.4 includes a small benchmark for each VP to generate the minimum number of transactions that activate all different patterns of modules' communication and transaction types.

In the following, as a representative of analyzing a SystemC TLM-2.0 VP, the result of analyzing the *AT-example* VP obtained by both proposed approaches is presented in detail.

AT-Example

The *AT-example* VP uses eight of the 13 TLM-2.0 base protocol's specified transaction protocols. The VP includes multiple approximately timed (hence the name of the model) initiators and targets, as well as an AT interconnect. The architecture of the *AT-example* design is shown in Fig. 3.23. It includes nineteen modules: four initiators, one interconnect, five targets, and nine checkers. It consists of two different types of initiator named *AT-typeA-initiator* and *AT-typeB-initiator*, and five different targets named *AT-typeA-target* through *AT-typeE-target*. Initiators (type *A* and *B*) and targets (type *A-E*) each implement different cases of a TLM communication's phase transitions (from the TLM-2.0 standard). For each communication that is done by each TLM module, the transaction packet is checked by the checker modules (which are *BP-chkr-init0* to *BP-chkr-trgt4*).

Table 3.5 illustrates the result of analyzing the *AT-example* using the proposed approaches. It shows that the VP has 12 different communication flows, eight different transaction types, and overall, 14 unique patterns of flow and type. Thus, 14 UML diagrams are generated to present the simulation behavior of the VP. This helps designers to understand how different modules communicate together and which type of the base protocol transactions is implemented in a given TLM-2.0 model. Therefore, instead of reading more than 3000 lines of code of the VP distributed over 11 files, a simple glance over the quickly generated UML diagrams can significantly facilitate the design understanding and analysis process.

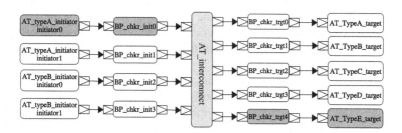

Fig. 3.23 The architecture of the *AT_example* VP

Table 3.5 The transaction classification result of the *AT_example* VP

Number	Flow	Modules	Transaction type
1	F_1	initA-init0→AT_intc→trgE-ins0	T_1
2	F_1	initA-init0→AT_intc→trgE-ins0	T_2
3	F_2	initA-init0→AT_intc→trgD-ins0	T_2
4	F_3	initA-init1→AT_intc→trgC-ins0	T_9
5	F_4	initA-init1→AT_intc→trgE-ins0	T_1
6	F_5	initA-init1→AT_intc→trgD-ins0	T_2
7	F_6	initB-init0→AT_intc→trgE-ins0	T_8
8	F_7	initB-init0→AT_intc→trgB-ins0	T_5
9	F_7	initB-init0→AT_intc→trgB-ins0	T_6
10	F_8	initB-init1→AT_intc→trgA-ins0	T_{11}
11	F_9	initB-init1→AT_intc→trgD-ins0	T_{13}
12	F_{10}	initB-init1→AT_intc→trgE-ins0	T_8
13	F_{11}	initB-init1→AT_intc→trgB-ins0	T_5
14	F_{11}	initB-init1→AT_intc→trgB-ins0	T_6

initA-init0 AT_typeA_initiator-initiator0, *AT_intc* AT_interconnect, *trgE-ins0* AT_typeE_target-instance0

As an instance, Fig. 3.24 illustrates a part of behavioral information of the *AT-example* (i.e., the gray component in Fig. 3.23) for a single transaction in the shape of a UML model. For each interaction between two modules, the corresponding arrow indicates both,

- operations that are called on a module instance and
- transaction data.

In this figure, the black stadium shapes illustrate the root name and instance name of modules and global functions within the design. For global functions, the instance name is *NULL*. The role of each component is presented on top of the components' name. It separates TLM modules from global functions or modules (e.g., for monitoring a transaction) which are not supposed to be a main part of transaction flow concerning the TLM-2.0 base protocol but still necessary to describe the model's behavior. For each interaction between two modules, the information on each arrow indicates operations that are called on an instance and are drawn from the caller to the callee concerning the simulation time. In particular, for a call from a TLM module or related modules (e.g., for monitoring a transaction), it shows the number of sequence, the name of the caller function, the timing phase, the delay time related to the phase transition, and the return value of the callee (if available). Regarding a call from a global function, it shows the sequence number and the name of the caller function.

In addition to the transaction flow, the generated UML model shows detailed transaction data. The box under each arrow contains the transaction's attributes, which are passed as an argument from the caller to the callee. The white boxes indicate a transaction object locally created by an initiator module and passed as a

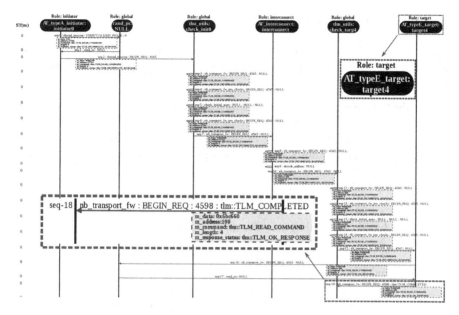

Fig. 3.24 A part of the generated UML model of the *AT-example* behavior presenting detailed transaction data

function argument, while the blue boxes show a transaction object reached the callee through a function call. As an example, *seq-18* in Fig. 3.24 shows the information about the response of the target module *AT-typeE-target*. It includes name of the called function (*nb_transport_fw*), timing phase of the transition (*BEGIN_REQ*), timing annotation (*4598 ps*), and the return value from *nb_transport_fw* (*tlm::TLM-_COMPLETED*). It also shows the transaction data passed to the callee module including the reference to the transaction data (*0x6bc660*), address (*100*), command (*tlm::TLM_READ_COMMAND*), length (*4*), and response status (*tlm::TLM_OK-_RESPONSE*). The obtained result in this example also demonstrates that the timing annotation that is used for the internal process does affect the simulation time, as it is expected from TLM transactions. Notice that the information also includes both, class names and instance names for the given modules, shown on the top of Fig. 3.24.

3.6.2 Integration and Discussion

In this section, we evaluate the characteristics and applicability of both proposed approaches on how they do meet the required criteria (which are discussed in Chaps. 1 and 2) that we expect an ESL VPs' design understanding approach satisfies them.

3.6.2.1 Characteristics Evaluation

As discussed in Chaps. 1 and 2 (Sect. 2.3.4), a design understanding (or analysis) approach must satisfy the following criteria to provide designers with a comprehensive solution.

1. Avoiding any pre-conditions concerning the source code or workflow in order to be as compatible as possible.
2. Applying as little intrusion to the existing sources and workflow as possible.
3. Extracting detailed information of a design which reflects its structure and behavior.
4. Supporting TLM constructs.

Hence, we evaluate the characteristics and the applicability of the proposed approaches based on how they meet the criteria mentioned above.

Concerning the First and Second Criteria

In comparison to approaches that rely on manipulating or modifying the SystemC library, interfaces or compilation workflow to extract the run-time information, both proposed approaches extract the detailed run-time information without any modification of the aforementioned standard workflow. In the case of the debugger-based approach, the information extraction process can be performed even without the source code being available as it only needs the binary executable model of the design and its debug symbols. It means that the proposed method is a non-intrusive and code-independent approach. In the case of the compiler-based approach, although the source code of a given VP is instrumented, the inserted instructions are a simple C++ instruction type *"Fout"* that may barley cause any compatibility issue during the design process. Moreover, the modification process is performed automatically, thus no manual effort is required. Hence, the proposed compiler-based approach does not inherit the main drawbacks of the traditional intrusive techniques that may cause the following problems in the design process.

- Compatibility issues with other applications in parallel during the design process that arise from modifying the SystemC library or its interfaces.
- The reduction of the automation degree due to expensive manual processes to manipulate the source code.
- A negative effect on the timing behavior or functionality of a given SystemC VP in comparison to the original model due to the SystemC kernel modification.

The proposed approaches can be combined with setups that already rely on a modified SystemC library, which makes the approach applicable to a wide range of ESL designs.

Regarding the Third and Fourth Criteria

Unlike the methods that focus on extracting the static aspect of an ESL implementation, the proposed approaches extract the VP models' behavior as well as their structure. Both of the proposed approaches can automatically retrieve a vast amount of information to describe the structure and behavior of a given ESL implementation. They present the retrieved information of a given ESL model in two different categories which are

- an XML model of the design architecture,
- a structured trace file of design's run-time information including

    - intra-cycle analysis of SystemC designs' behavior in the standard VCD format,
    - detailed information of transaction data and flow concerning the design's execution flow, and
    - transaction classification to eliminate and filter out the redundant transactions and generate UML presentation for all unique transactions w.r.t their lifetime.

3.6.2.2 Performance Evaluation

Tables 3.6 and 3.7 illustrate the execution time of both proposed approaches to analyze a given SystemC cycle-accurate and TLM-2.0 VP model, respectively. In both tables, the column *Compiler-based Approach* shows the required time of analyzing SystemC VPs using the compiler-based approach, while column *Debugger-based Approach* presents the execution time of the debugger-based approach. For both columns, *Phase1*, *Phase2*, and *VisT* illustrate the execution time of the first phase, the second phase, and the visualization phase of both proposed approaches, respectively. The *Total* column shows the total execution time, including all the phases above. Column *CET* demonstrates the total time (column *Total*) of each VP's compilation (column *Cmp*) and execution (column *Exe*) without any instrumentation. In the case of SystemC TLM-2.0 VPs, we compared the required time of analyzing a VP using the proposed approaches with the VP's pure compilation and execution time (column *CET*). In the case of SystemC cycle-accurate VP, in addition to the VP's pure compilation and execution time, the execution time of the proposed approaches was compared with the SystemC trace API (column *SC_API*) as well. As shown in Table 3.6, the SystemC trace API is the fastest approach to analyze the VP's behavior and has the lowest execution overhead in comparison to the VP's pure compilation time. However, the amount of extracted information using SystemC trace API is much less than the compiler-based and debugger-based approaches (demonstrated in Table 3.3). Furthermore, it needs manual effort by designers to insert the required instructions into the source code that increase its setup time.

Table 3.6 Experimental results related to the required analysis time for all SystemC cycle-accurate VP models

Test size	SystemC model	LoC	#Comp	#Test	SC_API[f] (s)	Compiler-based approach (s)				Debugger-based approach (s)				CET (s)		
						Phase1	Phase2[f]	VisT	Total	Phase1[f]	Phase2	VisT	Total	Cmp	Exe	Total
Small	4-bit shift-register[a]	135	2	50	0.01	1.46	0.02	0.11	1.59	1.26	1.41	0.21	2.88	2.18	0.01	2.19
	FIR-filter[b]	233	5	30	0.01	1.77	0.02	0.24	2.03	1.59	2.92	0.42	4.93	4.08	0.01	4.09
	3-stage-pipe[b]	290	5	10	0.01	1.81	0.02	0.26	2.09	1.73	3.66	0.41	5.80	4.81	0.01	4.82
	FFT_flpt[b]	484	3	10	0.01	2.73	0.02	0.58	3.33	3.11	46.15	1.16	50.52	3.86	0.01	3.87
	Cholesky[c]	522	2	27	0.01	2.82	0.02	0.81	3.65	3.96	11.49	1.24	16.79	4.32	0.01	4.33
	Hamming-code[d]	563	6	16	0.02	2.89	0.03	0.92	3.84	4.21	62.71	2.05	68.97	6.85	0.02	6.87
	Interpolation[c]	596	2	10	0.02	2.93	0.03	0.59	3.55	4.74	40.29	1.44	46.47	6.04	0.02	6.06
	IDCT[c]	815	2	100	0.02	3.11	0.03	1.24	4.38	4.32	26.37	2.19	32.88	3.51	0.02	3.53
	VGA[c]	856	6	100	0.02	4.72	0.03	3.90	8.65	6.90	249.21	7.52	263.63	3.74	0.02	3.76
	Decimation[c]	883	2	20	0.02	3.20	0.03	0.68	3.91	3.91	27.06	1.67	32.64	4.33	0.02	5.35
	pkt-switch[b]	1020	10	50	0.05	5.71	0.07	1.02	6.80	5.96	14.04	2.19	22.19	8.32	0.05	8.37
	MD5-hash[c]	1111	2	5	0.06	6.33	0.07	6.08	12.48	11.03	576.30	21.08	608.41	6.20	0.06	6.26
	RISC CPU[b]	1960	10	10	0.03	7.59	0.05	0.52	8.16	6.03	6.30	1.05	13.38	11.05	0.03	11.08
	Simple-bus[b]	2100	7	10	0.01	4.92	0.02	5.06	10.00	4.17	205.08	14.71	223.96	5.02	0.01	5.03
	AES128lowarea[c]	3280	13	5	1.40	12.23	1.47	14.37	28.07	16.01	694.19	32.08	742.28	16.03	1.39	17.42
	LZW-encoder[e]	5132	22	5	2.44	19.92	2.53	25.06	47.51	29.83	2683.20	46.11	2759.14	19.47	2.42	21.89

Large															
4-bit shift-register[a]	135	2	150	0.02	1.46	0.03	0.19	1.68	1.26	4.69	0.39	6.34	2.18	0.02	2.20
FIR-filter[b]	233	5	150	0.02	1.77	0.03	0.81	2.61	1.59	15.92	0.93	18.44	4.08	0.02	4.10
3-stage-pipe[b]	290	5	150	0.02	1.81	0.03	3.45	5.29	1.73	15.05	4.16	20.94	4.81	0.02	4.83
FFT_flpt[b]	484	3	150	0.02	2.73	0.03	6.2	8.96	3.11	233.18	13.02	249.31	3.86	0.02	3.88
Cholesky[c]	522	2	108	0.02	2.82	0.03	1.52	4.37	3.96	103.06	4.03	111.05	4.32	0.02	4.34
Hamming-code[d]	563	6	128	0.03	2.89	0.04	4.81	7.74	4.21	295.10	6.62	305.93	6.85	0.03	6.88
Interpolation[c]	596	2	104	0.03	2.93	0.04	4.69	7.66	4.74	191.01	15.39	211.14	6.04	0.03	6.07
IDCT[c]	815	2	996	0.04	3.11	0.05	5.42	8.58	4.32	228.00	19.05	251.37	3.51	0.04	3.55
VGA[c]	856	6	1000	0.04	4.72	0.05	11.12	15.89	6.90	877.05	44.71	927.66	3.74	0.04	3.78
Decimation[c]	883	2	215	0.03	3.20	0.04	3.72	6.96	3.91	345.92	15.22	365.05	4.33	0.03	5.36
pkt-switch[b]	1020	10	350	0.07	5.71	0.09	5.81	11.61	5.96	197.55	15.31	218.82	8.32	0.07	8.39
MD5-hash[d]	1111	2	50	0.07	6.33	0.07	16.17	22.57	11.03	2417.90	51.39	2480.32	6.20	0.08	6.27
RISC CPU[b]	1960	10	150	0.04	7.59	0.05	5.39	13.03	6.03	576.02	9.22	591.27	11.05	0.04	11.09
Simple-bus[b]	2100	7	100	0.02	4.92	0.03	11.59	16.54	4.17	1399.09	43.29	1444.55	5.02	0.02	5.04
AES128lowarea[c]	3280	13	50	3.41	12.23	3.49	22.33	38.05	16.01	3018.04	64.82	3098.87	16.03	3.39	19.42
LZW-encoder[e]	5132	22	50	5.18	19.92	5.41	31.44	56.77	29.83	11,498.30	93.51	11,621.64	19.47	5.16	24.63

LoC line of code, *#Comp* number of components, *SC_API* SystemC trace API, *VisT* visualization time, *CET* compilation and execution time, *Cmp* compile, *Exe* execution

[a] Provided by University of Edinburgh [54]
[b] Provided by OSCI [1]
[c] Provided by Schafer and Mahapatra [97]
[d] Provided by GitHub [30]
[e] Provided by [86]
[f] Please note that the reported time only includes the pure analysis and the compilation time is excluded

Compiler-Based Approach

As illustrated in Table 3.6, the total required time for analyzing a VP is almost the same or very close to the total time of its compilation and execution. The major timing-consuming part of this approach is the first phase where the AST of the VP is analyzed, and the instrumented version of the VP is generated. However, this time is still shorter than the VP's pure compilation in most cases. As illustrated in the table, this time is only related to the VP complexity and not the size of the running application (test cases). The execution time of the second phase of the compiler-based approach for each VP is very close to its pure execution time. It shows that although the generated instrumented VP model in the second phase of the compiler-based approach has more instruction than the original model, the overhead of executing the inserted instructions is negligible. The required time for the visualization phase depends on both the size of running application and the VP complexity. As illustrated in the table by increasing the design complexity and number of test cases, the extracted information also increases, thus the required time to visualize this information increases as well. This can be observed in the lower part of the table, where a large number of test cases are applied to each VP. For complex VPs such as *AES128lowarea* and *LZW-encoder*, the aforementioned fact is true even with a small number of test cases being applied to the VPs.

In the case of SystemC TLM-2.0 VPs, as demonstrated in Table 3.7, the first phase of the analysis for all VPs (regardless of the size of the running applications) is the major time-consuming part of the total execution time. The main reason is that the size of the generated log file, in this case, is less than the cycle-accurate model as only a limited set of variables (transaction objects) is traced and not all. However, for both types of VPs (cycle-accurate or TLM) increasing the design complexity and size of the running application (number of test cases or transactions) that needs to be extracted have direct effect on the total execution time of them. As shown in the table, by increasing the numbers of transactions (from small test size to large), the execution time increases as well. The first half of the table (with a small test size) illustrates that designs with more lines of code and AT timing model (which including more complex base protocol transactions) have larger execution time.

Debugger-Based Approach

As illustrated in both Tables 3.6 and 3.7, the total execution time for analyzing a VP, in most cases of small and some cases of extensive application, is still in a reasonable boundary in comparison to the total time of its compilation and execution. The major time-consuming part of this approach is the second phase, where the VP's run-time information is extracted under the control of GDB.

The time that is consumed to extract the run-time information depends on the three following parameters.

Table 3.7 Experimental results related to the required analysis time for all SystemC TLM-2.0 VP models

Test size	TLM model[a]	LoC	#Comps	TM	#Trans	Compiler-based approach (s)				Debugger-based approach (s)				CET (s)		
						Phase1	Phase2[b]	VisT	Total	Phase1[b]	Phase2	VisT	Total	Cmp	Exe	Total
Small	LT-example[a]	175	2	LT	16	1.12	0.06	0.11	1.29	1.93	5.91	0.29	8.13	1.02	0.06	1.08
	Routing-model[a]	456	6	LT	10	2.11	0.08	0.08	2.27	2.76	17.33	1.21	21.30	1.42	0.08	1.50
	Example-4[a]	547	2	AT	10	2.17	0.12	0.14	2.43	5.81	81.20	1.39	88.40	1.33	0.08	1.41
	Example-5[a]	650	7	LT	10	3.21	0.08	0.12	3.41	6.11	89.24	1.28	96.63	2.01	0.08	2.09
	Example-6[a]	713	9	AT	20	4.79	0.13	0.19	5.11	7.59	131.17	1.72	140.48	2.02	0.13	2.15
	AT-example[a]	3410	19	AT	20	19.05	0.16	0.26	19.47	22.41	452.53	1.81	476.75	20.03	0.16	17.19
	Locking-two[a]	4690	23	LT/AT	20	25.62	0.20	0.33	26.15	29.08	582.11	1.94	613.13	22.32	0.20	22.52
	SoCRocket[b]	50,000	20	LT/AT	100	52.82	0.64	0.95	54.41	146.39	2799.38	3.82	2949.59	26.72	0.64	27.36
Large	LT-example[a]	175	2	LT	160	1.12	0.11	0.39	1.62	1.93	122.18	1.66	125.17	1.02	0.10	1.12
	Routing-model[a]	456	6	LT	100	2.11	0.12	0.31	2.54	2.76	207.29	2.81	212.86	1.42	0.01	1.53
	Example-4[a]	547	2	AT	348	2.17	0.21	0.52	2.90	5.81	3439.08	16.63	3461.52	1.33	0.18	1.51
	Example-5[a]	650	7	LT	69	3.21	0.10	0.19	3.50	6.11	1718.92	2.69	1727.72	2.01	0.09	2.09
	Example-6[a]	713	9	AT	245	4.79	0.41	0.63	5.83	7.59	2903.72	11.09	2922.40	2.02	0.33	2.55
	AT-example[a]	3410	19	AT	49	19.05	0.24	0.39	19.68	22.41	1591.07	4.96	1618.44	20.03	0.34	17.19
	Locking-two[a]	4690	23	LT/AT	371	25.62	0.79	0.76	27.17	29.08	4639.83	17.24	4686.15	22.32	0.66	22.98
	SoCRocket[e]	50,000	20	LT/AT	1000	52.82	1.21	1.63	55.66	146.39	17,446.19	29.12	17,621.70	26.72	1.08	27.80

LoC line of code, *TM* timing model, *#Trans* number of transaction

[a]Please note that the VP models are modified to support more variety of transactions type and flow than the original models

[b]Please note that the reported time only includes the pure analysis and the compilation time is excluded

1. The complexity of each ESL design.
2. The amount of information that needs to be extracted.
3. The size of running application (test cases).

Concerning the first parameter, as the program is executed on GDB to store the state of the program during its simulation time, the execution has to be halted repeatedly. The number and duration of halting a program are very related to the instruction types such as loop, *wait* statements, and function call. Therefore, more complexity leads to increased execution time.

In order to show how the performance of the proposed approach is related to the second and third parameters, we evaluate it by defining two depths of data extraction specified by *Abstract* and *Detailed*, and two size of applications to be run on each design which are *Small* and *Large*. The *Abstract* data extraction depth includes all structural information and the sequence of all components activation. In addition to the information retrieved in *Abstract* level, the *Detailed* depth of data extraction contains all value changes of modules and functions variables during execution (i.e., comprehensive information about a given VP's behavior as illustrated in Tables 3.3 and 3.4). The aforementioned depths of data extraction can be specified in the first phase of the debugger-based approach using the designer interaction options introduced in Sect. 3.4.3 of this chapter.

The combination of both the information extraction depths and the size of the running applications provides designers with four corners of the ESL design analysis. As illustrated in Fig. 3.25, these corner cases are *Small-Abstract* (SA), *Small-Details* (SD), *Large-Abstract* (LA), and *Large-Details* (LD). The *SA* corner shows the fastest analysis and the least amount of information about the design, while the *LD* corner represents the slowest and most precise analysis on a given ESL design. As an example, in the case of SystemC cycle-accrue VP models, Figs. 3.26 and 3.27 illustrate the execution time of the debugger-based approach for two depths of data extraction of each design where a small and large set of tests are applied to them, respectively. These figures show that the required time to analyze SystemC VPs using the proposed approach in the SA, SD, and most cases of the SL corner is within a reasonable time frame (order of minutes) and boundary in comparison to its pure compilation and execution time. However, for some cases in the LD corner

Fig. 3.25 Information
extraction corner cases

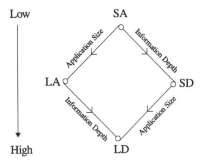

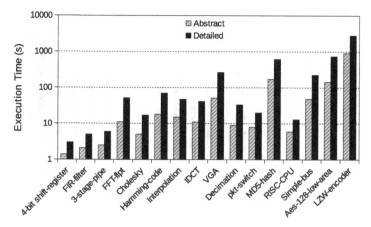

Fig. 3.26 Run-time analysis of the proposed debugger-based approach related to two depths of information extraction for small applications running on each SystemC design

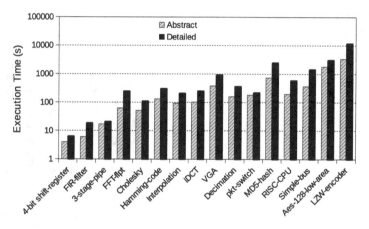

Fig. 3.27 Run-time analysis of the proposed debugger-based approach related to two depths of information extraction for large applications running on each SystemC design

(intricate designs such as LZW-encoder with a large amount of information needs to be extracted) the results in both Tables 3.6 and 3.7 show that the execution time is in order of hours.

In the following, two solutions are provided to overcome the aforementioned problem, which are caching data on the main memory instead of hard disk and parallelization. Since the major bottleneck of the debugger-based approach is the second phase of its analysis, where the run-time information is extracted under control of GDB, these implementation techniques should lead to an execution time reduction.

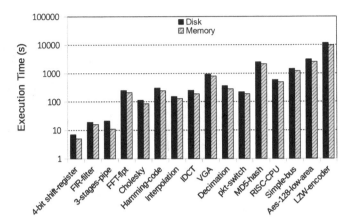

Fig. 3.28 The execution time of the debugger-based approach for analyzing SystemC cycle-accurate VPs for the LD corner running on disk versus memory

Fig. 3.29 The execution time of the proposed debugger-based approach for analyzing SystemC TLM-2.0 VPs for the LD corner running on disk versus memory

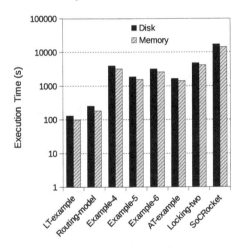

Caching Data on the Main Memory

This solution reduces the required time of storing the program states when executing it on GDB. We evaluated the execution time of analyzing SystemC cycle-accurate and TLM-2.0 VPs using the debugger-based approach for the LD corner. As illustrated in Figs. 3.28 and 3.29, storing data in memory instead of disk provides designers on average with 21% and 19% performance gain for SystemC cycle-accurate and TLM-2.0 VPs, respectively. It also shows that the performance of the proposed approach mostly depends on the complexity of each design. However, in the case of designs (e.g., *3-stage-pipe*) with simple structures (i.e., type of instruction), caching data in memory leads to 50% performance improvement.

Parallelization

This solution comes with the idea that an execution task can be broken down into several sub-tasks (which are similar) and processed independently. The obtained results are combined afterward whenever the execution of all tasks is completed. In our case, the task is to extract information about each module of the design, while the program is running on GDB. In the typical case, a script file is generated in the first phase of the debugger-based approach to program GDB to extract information related to all modules of the design during the simulation time on a single thread (as GDB does not support multi-threading to debug a program). Thus, to provide parallelism, the first phase of the proposed approach needs to be modified. Instead of generating a single script to program GDB, for each module of the design, one script file is generated. Thus, the task of information extraction can be run in parallel on different threads wherein each thread information related to a single module is extracted. The extracted run-time information are merged into a single log file whenever the execution of all parallel tasks is completed.

In the current implementation, we consider eight threads as the maximum degree of concurrency, meaning that at the same time, eight threads can be run on a system. Hence, for a design with more than eight modules, the information extraction task is distributed in such a way that all threads have the same workload. This is performed by calculating the complexity of modules (e.g., in terms of the number of functions and variables) and assigning them to the threads that at the end all have (almost) the same workload.

As shown in Figs. 3.30, and 3.31, the parallelization technique provides designers on average with 61% and 64% performance gain for SystemC cycle-accurate and TLM-2.0 VPs, respectively. The parallelization task is affected by two parameters,

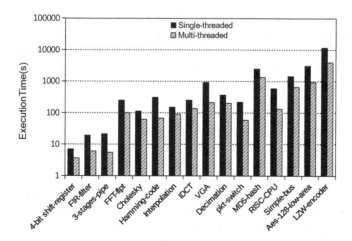

Fig. 3.30 The execution time of the proposed debugger-based approach for analyzing SystemC cycle-accurate VPs for the LD corner with and without parallelization technique

Fig. 3.31 The execution time of the proposed debugger-based approach for analyzing SystemC TLM-2.0 VPs for the LD corner with and without parallelization technique

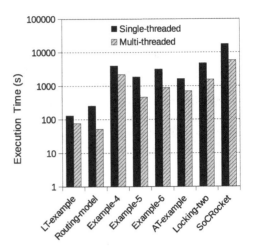

which are the number of VP's modules and complexity of modules. For a design with fewer modules than the maximum number of threads, the parallelization is very beneficial. The reason is that for design with more than eight modules, one thread must execute the information extraction task of more than one module. In this case, the effect of parallelization can be reduced as the required execution time for this thread may be more than the other seven threads. The performance gain can be up to (the number of threads)X as no threads need to wait for the completion of others. For example, the parallelization of the *3-stage-pipe* and the *Routing-model* VPs led to the execution time reduction of 74% and 79%, respectively.

Overall, the parallelization technique can effectively reduce the required execution time of the debugger-based approach for both types of VP (cycle-accurate and TLM-2.0). In this regard, the execution time, even in the LD corner for both types of VP, lies in order of minutes. The required execution time can even be more reduced, e.g., by increasing the number of threads or breaking the parallelization task down to each function of a module (instead of each module). Moreover, a combination of both the techniques above (caching data on memory and parallelization) can also lead to a better performance gain.

3.6.2.3 Memory Usage Evaluation

Tables 3.8 and 3.9 show the size of the extracted information and generated output data sets. In both tables, columns *Run-time Log* and *XML* illustrate the size of the generated log file and extracted VP's structure, respectively. Column *VCD* in Table 3.8 presents the size of the generated VCD for each SystemC cycle-accurate VP. The UML size reported in Table 3.9 is the average size of the generated UML diagrams for each TLM design. In both tables, the largest output file is the *Run-time log* (which is generated in the second phase of the proposed approaches) as it includes the detailed data and the complete history of the execution. The debugger-

Table 3.8 Experimental results related to the size of extracted information and generated output data for all SystemC cycle-accurate VP models

Test size	SystemC model	LoC	#Comp	#Test	Run-time log (MB)		VCD (MB)	XML (MB)
					Debugger-bA	Compiler-bA		
Small	4-bit shift-register[a]	135	2	50	0.271	0.110	0.012	0.004
	FIR-filter[b]	233	5	30	1.412	0.593	0.063	0.007
	3-stage-pipe[b]	290	5	10	1.174	0.464	0.022	0.007
	FFT_flpt[b]	484	3	10	1.533	0.649	0.041	0.011
	Cholesky[c]	522	2	27	1.704	0.739	0.032	0.010
	Hamming-code[d]	563	6	16	2.130	0.854	0.045	0.016
	Interpolation[c]	596	2	10	1.557	0.619	0.039	0.006
	IDCT[c]	815	2	100	4.722	1.941	0.058	0.004
	VGA[c]	856	6	100	12.251	5.339	0.072	0.010
	Decimation[c]	883	2	20	1.453	0.329	0.054	0.008
	pkt-switch[b]	1020	10	50	4.117	1.708	0.651	0.036
	MD5-hash[c]	1111	2	5	18.431	8.319	0.041	0.011
	RISC CPU[b]	1960	10	10	1.069	0.441	0.103	0.026
	Simple-bus[b]	2100	7	10	13.493	4.820	0.213	0.011
	AES128lowarea[c]	3280	13	5	42.678	15.129	0.061	0.021
	LZW-encoder[e]	5132	22	5	91.711	33.109	0.239	0.052

(continued)

Table 3.8 (continued)

Test size	SystemC model	LoC	#Comp	#Test	Run-time log (MB)		VCD (MB)	XML (MB)
					Debugger-bA	Compiler-bA		
Large	4-bit shift-register[a]	135	2	150	0.762	0.309	0.031	0.004
	FIR-filter[b]	233	5	150	6.233	2.870	0.257	0.007
	3-stage-pipe[b]	290	5	150	16.260	6.188	0.314	0.007
	FFT_flpt[b]	484	3	150	20.127	8.409	0.542	0.011
	Cholesky[c]	522	2	108	6.175	2.105	0.102	0.010
	Hamming-code[d]	563	6	128	16.719	5.331	0.311	0.016
	Interpolation[c]	596	2	104	15.108	6.772	0.345	0.006
	IDCT[c]	815	2	996	43.400	16.019	0.513	0.004
	VGA[c]	856	6	1000	110.376	37.004	0.640	0.010
	Decimation[c]	883	2	215	12.173	5.190	0.409	0.008
	pkt-switch[b]	1020	10	350	26.056	15.141	3.620	0.036
	MD5-hash[d]	1111	2	50	165.279	61.093	0.192	0.011
	RISC CPU[b]	1960	10	150	13.390	5.081	0.268	0.026
	Simple-bus[b]	2100	7	100	129.960	49.705	2.039	0.011
	AES128lowarea[c]	3280	13	50	419.117	171.902	0.219	0.021
	LZW-encoder[e]	5132	22	50	904.270	318.083	0.722	0.052

LoC line of code, *#Comp* number of components, *SC_API* SystemC trace API, *VisT* visualization time, *CET* compilation and execution time, *Cmp* compile, *Exe* execution

[a]Provided by University of Edinburgh [54]
[b]Provided by OSCI [1]
[c]Provided by Schafer and Mahapatra [97]
[d]Provided by GitHub [30]
[e]Provided by orahyn [86]

Table 3.9 Experimental results related to the size of extracted information and generated output data for all SystemC TLM-2.0 VP models

Test size	TLM model[a]	LoC	#Comps	TM	#Trans	Run-time Log (MB)		UML (MB)	XML (MB)
						Debugger-bA	Compiler-bA		
Small	LT-example[1]	175	2	LT	16	0.724	0.330	0.007	0.002
	Routing-model[1]	456	6	LT	10	5.108	2.419	0.014	0.004
	Example-4[1]	547	2	AT	10	8.690	3.461	0.019	0.003
	Example-5[1]	650	7	LT	10	6.119	2.344	0.014	0.005
	Example-6[1]	713	9	AT	20	15.114	6.931	0.038	0.017
	AT-example[1]	3410	19	AT	20	17.075	7.203	0.042	0.036
	Locking-two[1]	4690	23	LT/AT	20	19.323	9.601	0.038	0.047
	SoCRocket[2]	50,000	20	LT/AT	100	186.650	70.226	0.055	0.345
Large	LT-example[1]	175	2	LT	160	6.390	2.830	0.007	0.002
	Routing-model[1]	456	6	LT	100	61.420	29.537	0.014	0.004
	Example-4[1]	547	2	AT	348	406.220	101.098	0.019	0.003
	Example-5[1]	650	7	LT	69	57.140	24.077	0.014	0.005
	Example-6[1]	713	9	AT	245	213.278	87.650	0.038	0.017
	AT-example[1]	3410	19	AT	49	51.635	23.019	0.042	0.036
	Locking-two[1]	4690	23	LT/AT	371	379.404	105.618	0.038	0.047
	SoCRocket[2]	50,000	20	LT/AT	1000	1602.011	394.176	0.055	0.345

LoC line of code, *TM* timing model, *#Trans* number of transaction, *#UFlow* number of unique transaction flow, *#UType* number of unique transaction type, *#UML* number of generated UML diagram

[a]Please note that the VP models are modified to support more variety of transactions type and flow than the original models

[1]Provided by Aynsley [6]

[2]Provided by Schuster et al. [101]

based approach has a bigger log file than the compiler-based approach because of the following two main reasons.

First, to extract the required information, the debugger-based approach needs to walk through each line of code, while the compiler-based approach visits only pre-defined specific points. Second, in the case of the debugger-based approach, some further information related to the execution state (e.g., information related to breakpoints) is also logged, while in the compiler-based approach only the data of *retrieving* statements (which are *Fout* instructions) is extracted.

A large part of the information in the *Run-time log* is used to track the useful data related to the behavior and structure of the design and thus filtered out during the translation process (in the third phase). The XML and UML files for all case studies remain in the order of KB. Using the filter options can reduce the size of output data sets (VCD, UML, and XML) even more, allowing the designer to have a better readability, focusing on the resulting logs at whatever issue is currently at hand.

In the case of the debugger-based approach, the reported size for all case studies is in order of MB with the worst-case *LZW-encoder* and *SoCRocket* in LD corner, which is about one and two GB, respectively. This allows designers to usually store the extracted data in the first phase of this approach (creating the *Run-time log*) in the main memory (instead of the hard disk) to reduce I/O overhead and to improve the performance.

3.6.2.4 Limitations

As both of the proposed approaches are based on run-time analysis for the task of information extraction, they inherit the same limitation. The run-time information extraction and the subsequent visualization tasks are very related to the running software or application to activate different features and functionalities of a given VP. However, it is common practice in the design process to keep the design coverage (e.g., functional or statement) monitored and high to ensure that it works as intended. Thus, a design mostly contains a testbench with high coverage (at least one test to activate each unique functionality of the VP). This is usually provided by designers either manually or by employing an automated test generator.

Regarding the compiler-based approach, the only requirement of the proposed approach is the availability of the source code. In the case of the debugger-based approach, the only precondition of the suggested approach is that the executable VPs must contain debugging information that is compatible with GDB. While GCC and Clang-LLVM are supported, Microsoft Visual Studio is currently not. Although our approach is an overall sound analysis, it does share the inherent performance limitations that come with most other run-time analysis tools build on GDB. However, using the implementation techniques which are introduced in this section (i.e., caching data in the main memory or parallelization) can reduce the performance loss effectively.

3.6.3 Summary

Overall, the introduced approaches are able to extract detailed information related to the structure and behavior of a given ESL VP. The performance evaluation of both approaches shows that the compiler-based approach is a fast solution that scales very well with the arbitrary complexity of a given VP and the size of its running application. For all case studies in both Tables 3.6 and 3.7, the required time for analyzing VPs is less than a minute. The debugger-based approach has a reasonable execution time for analyzing VPs with a small size of application running on them (most cases lie in the order of seconds) in comparison to their pure compilation and execution time. Although the performance impact of the approach can be noticeable when analyzing a large given ESL design (e.g., LZW-encoder or SoCRoket), there are a wide variety of use cases that justify the application of the proposed approach.

- First and foremost, the required setup time is negligible. Often, approaches that are introduced in Chap. 2 require a considerable amount of work until they work correctly. Any methods that alter the compilation workflow more often than not take hours (if not days) to set up until they work properly. The AOP solution that was referenced, e.g., seems to be a straightforward way to instrument the source code without altering the original code base, relying on automatic re-writing in an additional step that is executed automatically before compilation. However, when being applied in practice, setting up a working AOP environment for existing, more extensive projects, can quickly turn out a significant amount of work, with the aspect weaver first being required to parse the whole project (which is primarily an issue as soon as libraries are involved) and the resulting code often causes new issues for a given compiler setup. The presented solution aims to be a simple fire-and-forget approach, requiring only an executable with the debug symbols attached (which should be available for all projects). Due to relying on this standard workflow, the setup costs remain low and problematic corner cases may be traced before other setups are even running.
- The built pipeline remains untouched. This keeps the setup time low and lets designers quickly evaluate, especially short and questionable test cases. Moreover, any results obtained using the given method are sure to be identical to those that are shown for the production system.
- In the case of 3PIP for which only an executable binary is available, the proposed approach remains operable. It is the only solution that works on an executable model of the design without the availability of source code—retrieving the information that is available via debug symbols and omitting the unavailable data without further setup.

Two implementation techniques (i.e., caching data on memory and parallelization), which were introduced in this chapter, can be used by designers to gain better performance when using the debugger-based approach. The suggested techniques can reduce the required analysis time for the LD corner down to an order of minutes even for complex VPs.

3.7 Conclusion

In this chapter, first, two motivating examples were introduced to show the necessity of the SystemC-based VP analysis approach and design understanding at the ESL. These examples illustrated that the design understanding is the first step before any modification can be applied to a given VP and helpful for other tasks such as debugging in the design process.

Second, we indicated that to understand a given VP properly, what types of information need to be extracted, and how they must be visualized to empower designers to grasp the intricacy of VP quickly.

In this respect, a comprehensive design understanding methodology was introduced including (1) two new easy-to-use ESL VPs information extraction and analysis approaches which are based on the debugger (i.e., GDB) and compiler (i.e., Clang) and (2) a post-execution analysis on the extracted data to provide designers with an IR of the VPs' behavior (based on simulation) and structure, and visualize this information in such a way that designers can quickly grasp the complexity of VPs. Both VP information extraction approaches can retrieve structural and behavioral information of VPs implemented using either SystemC cycle-accurate or SystemC TLM-2.0 models.

The debugger-based approach requires only the executable version of the VP, thus the original source code and workflow (e.g., SystemC library or compiler) stay untouched. The main problem with intrusive approaches that rely on altering e.g., the SystemC library, interfaces, SystemC simulation kernel, or compiler is that these modifications may cause an issue for the application of several approaches in parallel, future updates or restrictive environments. Moreover, they mostly reduce the degree of automation as they require designers effort. However, the required time for this analysis might be expensive when the VP complexity increases.

On the other hand, the compiler-based approach is very fast and scales well with an arbitrary complexity of VPs. However, it requires the availability of the VP original source code. It works based on analyzing the AST of the VP to generate an instrumented version of the VP for data extraction purpose. Since the proposed approach modifies neither the SystemC library nor the SystemC simulation kernel nor compiler, any results obtained using the approach basically should be identical to the reference results in terms of VP's timing behavior and its functionality.

Overall, due to the designer's concerns and requirements, they have this option to choose either debugger-based or compiler-based approach. The retrieved information using both proposed approaches can be effectively utilized for ESL designs verification, validation, and design space exploration.

Chapter 4
Application I: Verification

In the previous chapter, two novel SystemC-based VP analysis approaches were introduced, allowing designers to access detailed information about both the structure and behavior of a given VP. Apart from the design understanding goal (which is one of the primary goals of this book), the results of this phase can also be employed as a foundation for various applications in the design process. This utilization is the second primary goal of this book, which aims to show how the extracted information can be used to facilitate performing other tasks in the design process.

In this chapter, we show one application of the design understanding in the design process, which is the verification of SystemC-based VPs implemented using the SystemC TLM-2.0 framework.

4.1 Introduction

The much earlier availability and the significantly faster simulation speed of the VP in comparison to the RTL hardware model are the main reasons that VP used as a reference model for an early system verification in the design process. Hence, ensuring the correctness of VPs is of the utmost importance, as undetected faults may propagate to lower levels of abstraction and become very costly to fix.

In the ESL design, TLM-2.0 (as the current standard) provides designers with a set of standard interfaces and rules (TLM-2.0 base protocol) to model a VP based on abstract communication (i.e., transactions). As mentioned earlier, it allows designers to abstract away the implementation details related to the computation of IPs and only focus on communication. Thus, communication (among different IP cores) is the main part of a VP model that must be verified. The first step to verify the communication in a given VP is to check whether or not they adhere to the TLM-2.0 rules. The TLM-2.0 standard comes with more than 150 rules that must be

© Springer Nature Switzerland AG 2020
M. Goli, R. Drechsler, *Automated Analysis of Virtual Prototypes at the Electronic System Level*, https://doi.org/10.1007/978-3-030-44282-8_4

adhered to when a TLM model is implemented [5] and which define the expected behavior of a TLM-2.0 compliant model. A small part of these rules specifies limitations concerning the structure of TLM models (e.g., an initiator socket must be connected to a target socket). This set is called the *static rules* and is checked during compilation. A large part of these rules (the *dynamic rules*) determines restrictions on the TLM communication and transaction attributes—and as such defines, a set of protocols that TLM designs must adhere to. Neither the SystemC compiler nor the TLM library detects TLM protocol violations that occur during execution. Manually verifying all rules and detecting the source of any given error is error-prone and expensive even for simple models and thus practically impossible for complex designs. Therefore, automated verification techniques that verify the compliance of a given ESL model with at least the base protocol are needed.

In addition, as a VP is the first executable model of the design specifications describing its functionality and timing behavior in terms of abstract communication, a functional assurance of the VP against its specifications is necessarily required, especially if the VP under development represents a safety-critical system. Therefore, to ensure the correctness of communication in a given VP, apart from validating the VP against TLM-2.0 rules (protocol validation), the functionality and timing behavior of the VP must be verified as well.

In general, the SystemC-based VP correctness can be ensured by two different approaches: formal verification and simulation-based verification (also called validation). Formal approaches usually require to specify the model in formal semantics such as abstract state machines [18, 45, 48, 51] or IR models [66]. However, due to the object-oriented nature and event-driven simulation semantics of SystemC, it is very challenging to verify a given SystemC VP formally. Moreover, the state-space explosion is another well-known problem for this category. Due to these restrictions, formal approaches are not able to verify complex systems as several assumptions need to be imposed on a given VP (e.g., function pointers, recursion, or templates cannot easily be described formally).

In contrast, simulation-based verification approaches [28, 38, 39, 58, 89, 104, 110] are still the predominant techniques to verify systems at the ESL as they scale very well with an arbitrary complexity of VPs. In simulation-based verification, the behavior of SystemC models is verified during the simulation. In this scope, assertion-based techniques [24, 28, 89, 104, 109] are particularly well-suited for validation purposes. However, they come with some major drawbacks as the following. First, deriving assertions (i.e., properties) from TLM-2.0 rules or the design specifications usually requires manual effort by designers. Second, the generated assertions mostly need to be inserted manually to the VP. Third, in many cases, the SystemC kernel [109] or the SystemC library [104] needs to be modified to trace transactions accurately. This either relies on expensive manual processes or cause compatibility issues that overall reduce the degree of automation.

In this chapter, we propose a simulation-based verification approach which automatically validates a given SystemC VP against both the TLM-2.0 rules and its specifications. We take advantage of the information extracted approaches introduced in Chap. 3 to access the run-time information of the VP. A post-execution analysis is applied to the extracted information to build the simulation behavior of the VP and to describe it in terms of abstract communication (i.e., transactions). The simulation behavior is transformed into a finite state machine, and properties are automatically generated from the design specifications and TLM-2.0 rules. Finally, the translated model (which formally describes the simulation behavior of the VP) is checked against the generated properties by a model checker. The violated properties and the corresponding transactions are reported back to the designers for further analysis.

The focus of the proposed approach is to detect the errors related to the most common and essential fault types of communication in a VP at the ESL; i.e., dynamic rules (that cannot be checked statically, e.g., during compilation time) related to the TLM-2.0 base protocol transactions (and its attributes), functionality, and timing behavior. The approach is applied to several case studies, including a real-world VP to demonstrate its precision, scalability, and advantages.

The rest of this chapter is organized as the following. In Sect. 4.2, the related works of the VP verification at the ESL are explained. Section 4.3 introduces the types of fault that we target to verify in this chapter using a motivating example. Section 4.4 presents the main contributions of this chapter. It introduces the proposed verification methodology, including how the simulation behavior of a given VP is translated into a set of state machines, and the TLM-2.0 rules as well as the VP's specifications into a set of properties. Moreover, it illustrates how the validation process is applied to the transformed models. The evaluation and experimental results are presented in Sect. 4.5. Finally, the chapter is concluded in Sect. 4.6.

4.2 Related Works

As mentioned in the previous section, the methods for verifying SystemC models are mostly divided into two main categories: formal and simulation-based methods.

4.2.1 Formal Methods

Many approaches [45, 48, 51, 52] have been introduced to verify SystemC models formally. They rely on translating the models to formal semantics such as state machines and verifying them by checking safety properties. In [48], SystemC models are transformed into the *Abstract State Machine Language* (AsmL) and properties are formulated using the *Property Specification Language* (PSL). The translated model is verified using a model checker. In [45], a SystemC-TLM design is translated to a sequential C model and the properties are defined using PSL. Then, a monitoring logic which is based on C assertions and finite state machines is utilized to verify the properties. In [51, 52], a transformation of a given SystemC design to a timed automata model is created. The translated model is checked by model checkers such as UPPAAL and BLAST. The method proposed in [18] formalizes the semantics of SystemC designs in terms of Kripke structures and verifies it using symbolic model checking. The authors of [66] present an approach to translate SystemC models to an *Intermediate Verification Language* (IVL). Then, they provide a symbolic simulator to verify the IVL.

For all of the aforementioned methods, translation of SystemC designs to formal semantics is the main challenge confining them to verify only a subset of SystemC models.

4.2.2   *Simulation-Based Methods*

Several works [25, 58, 110] have been proposed to verify SystemC models using the AOP technique. The idea of AOP is based on a source-to-source translation enabling designers to add additional code (aspects) at specific points of the source code. In [58], the AOP technique is utilized to instrument the source code of SystemC-TLM models to trace its simulation. The traced information is checked against the properties implemented as a C++ class. In [25], SystemC source code and properties written as aspect description in XML format by users are parsed. A weaver module is used to reassemble the parsed source code and aspects. Then, a SystemC formatter module is utilized to generate a SystemC source code from the woven syntax tree to check the properties during the simulation time. In [110], user code primitives are defined in property specifications by users. Then, AOP is used to instrument the SystemC source code by generating a monitor for each property to be checked during the execution.

The AOP-based methods mostly depend on user interaction to define the aspects and design primitives. Additionally, the difficulty of defining and debugging AOP setups makes the methods rather hard to be used.

There are also several attempts [24, 28, 104] to verify SystemC models by *Assertion-based Verification* (ABV). In ABV, properties are specified as assertions (written in languages like PSL or System Verilog) and checked during simulation time. However, the methods have several disadvantages. Property formulation requires complex specification which is done manually and thus challenging to be applied during the simulation time. Mostly, it requires manual effort by designers to insert assertions into the design's sources or modify the SystemC constructs.

In [6, 7], SystemC-TLM models are verified by adding some TLM protocol checkers as an external SystemC module to monitor transactions. Transactions are monitored during the simulation by inserting a copy of the protocol checker between every pair of TLM modules. They check whether or not each transaction satisfies the TLM protocols. However, inserting these protocol checkers and external modules require manual effort by designers, which reduces the degree of automation. Furthermore, the method can only check the validity of a given VP against TLM-2.0 rules and not its specifications.

4.2.3 Summary

The existing methods have two major drawbacks: in case of formal verification, methods are limited to only support a subset of ESL designs. Regarding simulation-based verification, they are usually using intrusive techniques, either altering the SystemC library or kernel or its interfaces. This can reduce the degree of automation or create compatibility problems for the application of several approaches in parallel. Moreover, they are mostly restricted to verify only SystemC TLM-2.0 designs against TLM-2.0 rules.

4.3 Motivating Example and Fault Types

In this section, we introduce the types of fault that we target to detect in a given VP using a motivating example.

Consider the *LT_AT_BUS* VP (Fig. 3.1) presented in Chap. 3 but with different specifications as the following. Module *Initiator_A* communicates with target modules through *LT_AT_BUS* by generating three types of AT transactions. It generates transaction types T_1 and T_2 (w.r.t Table. 2.1) to access *Memory_A* (each type for different memory address ranges), and type T_3 to access *Memory_B* and *Memory_C*. The *Initiator_B* module generates AT transactions type T_4 to access target modules *Memory_A*, and the LT transactions type T_0 to access *Memory_B* and *Memory_C*. For example, consider the communication between *Initiator_A* and *Memory_A*. The *Initiator_A* module generates transactions of types T_1 (including fewer transition phases to gain performance) and T_2 to access memory address range ($0x00$ to $0x0A$) and ($0x0B$ to $0xFF$) of the *Memory_A* module, respectively. Now consider three possible *Fault Types* (FTs) that designers may face during the design process.

FT1 Implementing an incorrect TLM-2.0 base protocol transaction. After implementing the VP, some TLM-2.0 based protocol rules might be implemented incorrectly by designers. For example, the generated transactions by *Initiator_A* to access address range ($0x0B$ to $0xFF$) of target module *Memory_A* have wrong transition phase orders. However, this type of faults cannot be detected by either SystemC compiler or TLM-2.0 library. Moreover, checking the correctness of these rules manually, even for a simple design, is very difficult as TLM-2.0 includes many rules.

FT2 Transporting incorrect transaction data can also cause malfunction for the VP model. Transaction's attributes such as data, address, or data length can be assigned to a wrong value by each of the TLM modules, resulting in transporting transaction to an incorrect destination or receiving a wrong data by the initiator or target modules. For example, some transactions of type T_1 generated by the *Initiator_A* access the memory addresses $0xB2$ and $0xB3$, which are against the VP specifications. In this case, although the VP implementation adheres the TLM-2.0 rules, it violates the VP specifications. Thus, the traditional verification approaches such as [104] that only focus on verifying the VP against the TLM-2.0 base protocol rules, fail to detect this type of fault.

FT3 Implementing an incorrect timing behavior can also cause an error. This is explicitly related to the timing annotation of a transaction and defined as the delay that the transaction requires to be transferred between two TLM modules. For example, the delay parameter of transactions generated by *Initiator_A* to access the address range ($0x00$ to $0x0A$) of *Memory_A* must be less than 80 ns. Hence, designers want to know whether or not the timing behavior of the generated transactions by *Initiator_A* to access this range of address in *Memory_A* adheres the VP's timing specifications. For this type of faults, again, traditional approaches such as [104] fail to validate the VP against its specifications.

Hence, to detect all the aforementioned types of fault, an automated validation process is required that not only checks the correctness of VPs against TLM-2.0 rules but also validates their functional and timing behavior against their specifications.

4.4   Methodology

In this section, we first give an overview of the proposed verification approach. Second, the process of extracting the simulation behavior of VPs and translating it into the proper formal format for the task of verification is presented. Then, we show how the TLM-2.0 rules and VPs' specifications are translated into the formal semantics. Finally, the verification of the formal presentation of VPs' simulation behavior against the generated properties is illustrated [36, 38, 39].

4.4.1   Overall Workflow

Figure 4.1 provides an overview of the proposed approach which includes three main phases as below.

1. Extracting the run-time behavior (i.e., transactions' lifetime) of the SystemC VP and analyzing the extracted information to transform it into a set of timed automata models.
2. Generating a set of properties from the following sources.

   - TLM-2.0 based protocol rules and
   - VP specifications related to its functional and timing behavior.

3. Validating the VP's behavior against the generated properties using the UPPAAL model checker.

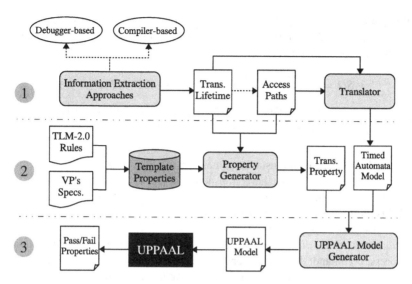

Fig. 4.1 The proposed simulation-based verification methodology overview

In the following, each phase of the proposed approach is explained in detail and illustrated using the motivating example, *LT_AT_BUS* VP (Fig. 3.1).

4.4.2 Information Extraction and Translation

The VP's behavior extraction in the first phase of the verification approach is performed using the suggested information extraction approaches in Chap. 3. This option is available for designers to use either the compiler-based or debugger-based approach based on their use cases and requirements. One of the intermediate representation of the VP's simulation behavior using the design understanding approaches is the transactions' lifetime log. For the design understanding intention, the generated log includes detailed information related to the TLM modules and global functions or modules behavior (e.g., for monitoring a transaction). Some parts of this information, however, are necessary for the design understanding intention (as supposed to describe the whole behavior of the VP), they are not the main part of transaction flow concerning the TLM-2.0 protocols and rules. Hence, the generated transaction lifetime *Trans Lifetime* in the first phase of the proposed approach (Fig. 4.1) only includes the required information for the task of verification. Thus, each transaction lifetime in *Trans Lifetime* includes several sequences (timing steps), illustrating the transaction creation, manipulation by TLM modules, and its completion. Each sequence in the transaction lifetime is defined based on the following definition.

Definition 4.1 A sequence SQ is a tuple (F, D) where F is all information related to the transaction's flow and D denotes the information describing the transaction's attributes and its related parameters.

$$SQ = \{(F, D) \mid F = (M, I, Func, ST, TID, MT),$$
$$D = (data, adrs, cmd, dl, rps, phase, delay, rs)\}$$

where

- M and I are the root and instance names of a module, respectively.
- $Func$ is the function name that a transaction object is referred.
- ST shows the simulation time.
- TID is the transaction reference address.
- MT illustrates the type of TLM modules.
- $data$, $adrs$, cmd, dl, and rps are the transaction's attributes denoting the data, address, command, data length, and the response statues, respectively.
- $phase$, $delay$, and rs present the transaction's phase, timing annotation, and return status of the communication interfaces, respectively.

Please note that the *phase* parameter in the LT model is set to *NULL* as it is only relevant to the AT model. Based on Definition 4.1, transaction lifetime is defined as the following:

Definition 4.2 A transaction lifetime TL is a set of sequences SQ where

$$TL = \{SQ_i \mid 1 \leq i \leq n_T\}$$

and n_T is defined based on which base protocol transaction is used as different types have disparate number of sequences.

For example, Fig. 4.2 illustrates a single transaction lifetime of the LT_AT_BUS VP based on Definitions 4.1 and 4.2, including five sequences.

Although the generated transaction lifetime TL has the proper structure to be used in the validation of each transaction against the TLM-2.0 base protocol rules, further translation on the transaction lifetime is required to verify functional (FT2) and timing (FT3) fault types. Since the validation of a given VP's transactions against FT2 or FT3 requires to check whether or not the transactions are sent to the right target module (e.g., a right memory address) with the expected transaction type or delay w.r.t the VP specifications, we transform each transaction lifetime into an access path based on the following definition.

Definition 4.3 A complete simulation behavior of a given SystemC VP can be defined as a set of access paths SAP where each path AP shows a connection between an initiator module IM and a target module TM as below

$$SAP = \{AP_i \mid AP_i = \{IM \rightarrow TM :: (TID, TT, Tadrs, cmd, TD)\}, \quad 1 \leq i \leq n_{seq}\}$$

SQ1: ([Initiator_A, init_0, process_1, 50ns, NULL, initiator],
[0x7054f0, 0x06, READ, 4, TLM_INCOMPLETE_RESPONSE, BEGIN_REQ,
5ns, NULL])

SQ2: ([LT_AT_BUS, bus_0, nb_transport_fw, 55ns, 0x683c10,
interconnect], [0x7054f0, 0x06, READ, 4, TLM_INCOMPLETE_RESPONSE,
BEGIN_REQ, 5ns, NULL])

SQ3: ([Memory_A, trgA, nb_transport_fw, 60ns, 0x683c10, target],
[0x76ab56, 0x03, READ, 4, TLM_OK_RESPONSE, BEGIN_REQ,
5ns, TLM_COMPLETED])

SQ4: ([LT_AT_BUS, bus_0, nb_transport_fw, 65ns, 0x683c10,
interconnect], [0x76ab56, 0x03, READ, 4, TLM_OK_RESPONSE,
BEGIN_REQ, 5ns, TLM_COMPLETED])

SQ5: ([Initiator_A, init_0, process_1, 70ns, NULL, initiator],
[0x76ab56, 0x03, READ, 4, TLM_OK_RESPONSE, BEGIN_REQ,
5ns, NULL])

Fig. 4.2 A single transaction lifetime of the *LT_AT_BUS* VP w.r.t Definitions 4.1 and 4.2

where

- *IM* and *TM* are initiator and target modules (their root and instance names), respectively.
- *TT* is the transaction type illustrating which timing model (LT or AT) is used. In the case of the AT model, it shows which type of the based protocol transaction is implemented w.r.t Table 2.1. To identify the transaction type, a unique type signature is generated by concatenating three parameters from the transaction lifetime, which are *communication interface call*, *return status*, and *phase transitions*.
- *Tadrs* shows the address of the transaction in the target module *TM*.
- *cmd* is the transaction command attribute. It shows the type of the transaction access (e.g., read or write).
- *TD* is the total delay required for a transaction lifetime completion. This is obtained by differentiating the simulation time *ST* of the first and last sequences.
- n_{seq} is the number of sequence in a transaction lifetime.

For example, the transaction lifetime presented in Fig. 4.2 has five sequences and implement type T_1 of the based protocol transaction as its type signature is "*nb_transport_fw+BRQ+TC*." The access path representation of this transaction lifetime based on Definition 4.3 is as below.

$$AP = \{Initiator\_A : init\_0 \rightarrow Memory\_A : trgA :: \qquad (4.1)$$

$$(0x683c10, T_1, 0x03, READ, 20ns)\}$$

It shows that the instance *init_0* of initiator module *Initiator_A* created a transaction with reference address *0x683c10* to write in memory address *0x03* of the instance *trgA* of target module *Memory_A*. It also indicates that the overall delay for this transaction is *20* ns as its first (SQ_1) and last (SQ_5) sequences are started at simulation time 50 ns and 70 ns, respectively.

4.4.2.1 Translating Transaction Information into Timed Automata.

In order to formally verify the simulation behavior of TLM models, it is necessary to transform the retrieved information into a formal model. As the extracted transactions' lifetimes have finite steps to implement the TLM protocol and include timing information (e.g., timing phase and timing annotations), they can be transformed into a timed automata model. A timed automaton is a finite state machine controlled by clock variables.

The following definitions are used to transform a transaction lifetime into a timed automata model.

Definition 4.4 A timed automaton *TA* is a tuple (L, l_0, C, A, E, I), where L is a set of locations, $l_0 \in L$ is the initial location l_0, C is a set of clock variables, and A is a set of actions, $E \subseteq L \times A \times B$ $(C) \times 2C \times L$ is a set of edges, where $B(C)$ denotes a set of clock constraints, and $I : L \to B(C)$ assigns invariants to locations. The transition $l \xrightarrow{(a,g,r)} l'$ is valid when $(l, a, g, r, l') \in E$.

Definition 4.5 The semantics of a *TA* is defined as a transition system $(S, s_0, \to)$, where $S \subseteq L \times R_{\geq 0}^{|C|}$ is a set of states $s_0 = (l_0, u_0)$ the initial state and $\to \subseteq S \times (R_{\geq 0} \cup A) \times S$ the transition relation. A clock valuation is a function $u : C \to R_{\geq 0}$ that maps a non-negative real value to each clock. A semantic step of a timed automaton to model the simulation behavior is defined as

$$(l, u) \xrightarrow{a} (l', u') \text{ iff } l \xrightarrow{(a,g,r)} l' \text{ such that } u \in g \wedge u' = [r \to 0]u \wedge u' \in I(l').$$

A transaction lifetime includes several timing steps which present the transaction creation and manipulation by TLM modules. In order to transform the transaction lifetime into a timed automaton, the following two steps are required. First, each sequence is defined as a state (location). Second, the set of data exchanged between modules is defined as the transition of states (edges). A location is defined based on the information of the module and the sequence number. In this model, a clock variable is used to control the sequence of transitions. A location is defined based on the information of the module and the sequence number. In this model, a clock variable is used to control the sequence of transitions.

A simple example illustrates how the extracted simulation behavior of a TLM model is transformed into a timed automata model. Figure 4.2 illustrates a part of the simulation behavior of the *LT_AT_BUS* VP (Fig. 3.1), which is the lifetime of a single transaction. A transaction is created by the initiator module *Initiator_A* with phase *BEGIN_REQ* and passed through the interconnect module *LT_AT_BUS* to reach the target module *Memory_A*. The target module returns *TLM_COMPLETED* when it receives the transaction.

The timed automaton *TA_0x7054f0_1* formally denotes the simulation behavior of Fig. 4.2 based on Definition 4.4 as below:

$$L = \{l_0 : SEQ1\_Initiator\_A,\ l_1 : SEQ2\_LT\_AT\_BUS, \tag{4.2}$$

$$l_2 : SEQ3\_Memory\_A,\ l_3 : SEQ4\_LT\_AT\_BUS,$$

$$l_4 : SEQ5\_AT\_typeA\_initiator\}$$

$$l_0 = SEQ1\_Initiator\_A$$

$$C = \{clk\}$$

$$A = \phi$$

$$E = \{(l0, l1),\ (l1, l2),\ (l2, l3),\ (l3, l4)\}$$

$$I : l_1 \rightarrow clk \leq 1,\ l_2 \rightarrow clk \leq 2,\ l_3 \rightarrow clk \leq 3,\ l4 \rightarrow clk \leq 4$$

The clock variable *clk* is initialized to zero and then used in two clock conditions. First, the invariant $clk \leq maxtime$ indicates that the corresponding location must be left before *clk* becomes greater than *maxtime*, and the guard $clk == maxtime$ enables the corresponding edge at *mintime*. $A = \phi$ denotes that all transitions between locations are the internal transitions. It means that a transition is taken if only the guard condition of the edge is satisfied.

The operational semantics of the timed automaton *TA_0x7054f0_1* is formulated based on Definition 4.5 as below:

$$(l_0, clk \leq 0) \xrightarrow{AS_0, clk==0} (l_1, clk \leq 1),\ (l_1, clk \leq 1) \xrightarrow{AS_1, clk==1} (l_2, clk \leq 2) \tag{4.3}$$

$$(l_2, clk \leq 2) \xrightarrow{AS_2, clk==2} (l_3, clk \leq 3),\ (l_3, clk \leq 3) \xrightarrow{AS_3, clk==3} (l_4, clk \leq 4)$$

The parameter $\overset{3}{\underset{i:0}{}} AS_i$ is defined as an assignment statement when a transition is taken from l_i to l_{i+1}. It sets the value of transaction attributes and the value of corresponding variables describing the transaction flow.

The access path representation of a transaction lifetime is a simplified model of the transaction lifetime that only has one transition phase. Thus, its timed automaton model only contains two locations l_0 and l_1, where the former refers to the initiator module and the later the target module. For example, the timed automaton *TA_0x7054f0_1_AP* formally denotes the access path in 4.1 (of Fig. 4.2) based on Definition 4.4 as the following:

$$L = \{l_0 : SEQ1\_Initiator\_A, \; l_1 : SEQ2\_Memory\_A\} \tag{4.4}$$

$$l_0 = SEQ1\_Initiator\_A$$

$$C = \{clk\}$$

$$A = \phi$$

$$E = \{(l0, l1)\}$$

$$I : l_1 \rightarrow clk \leq 1$$

The operational semantics of the timed automaton $TA\_0x7054f0\_1\_AP$ is formulated based on Definition 4.5 as the following:

$$(l_0, clk \leq 0) \xrightarrow{AS_0, clk==0} (l_1, clk \leq 1) \tag{4.5}$$

The parameter AS_0 is defined as an assignment statement when a transition is taken from l_0 to l_1. It sets the transaction's type, attributes (which is data, address, and command), and the duration of its lifetime (considered as the transaction delay).

4.4.3 Property Generation

Like the simulation behavior of TLM models, the TLM-2.0 rules and the VP's specifications need to be formally expressed in a well-defined language in order to compare the former to the latter. Properties are specified using a subset of *Timed Computation Tree Logic* (TCTL) [3] and described in the language of temporal logic. The language includes state formulas and path formulas. State formulas are expressions that are checked for a state, while path formulas evaluate whether a given state formula is satisfied over paths by any reachable state.

We take advantage of the following pre-defined symbols, which are taken from the field of temporal logic to formally define both the TLM-2.0 rules and the VP's specifications.

Definition 4.6 If p and q are a property of states, then the temporal logic formula:

- **Exists eventually** p ($E <> p$) describes that there is a path that leads to a state in which p holds.
- **Exists globally** p ($E [ ] p$) describes that there is a path in which p holds for all the states of the path.
- **Always globally** p ($A [ ] p$) describes that p holds for all states of all paths.
- **Always eventually** p ($A <> p$) describes that all paths p hold for at least one state of the path.

- q **always leads to** p ($q \rightarrow p$) describes any path that starts with a state in which q holds later reaches a state in which p holds.

4.4.3.1   Translating TLM-2.0 Rules into Formal Properties

As illustrated in Fig. 4.1, phase 2, TLM rules are transformed from the textbook specifications written in the TLM documentation into a set of template properties. This process is done manually (once) to create a database of pre-defined temporal properties. TLM rules are defined formally using pre-defined symbols, which are taken from the field of temporal logic using Definitions 4.6.

The *Template properties* database contains the properties to verify both the transaction semantics (i.e., TLM module behavior, the transaction types, and its attributes) and the functionality of TLM communication (i.e., TLM modules behavior). There are 40 template properties, of which 15 are defined to verify the semantics of a transaction and 25 are used to check the compliance of the semantics of the communication against TLM-2.0 rules.

As an example of *transaction semantics* rules, the TLM rules related to the default value of a transaction address attribute are considered. The textbook specification of this rule in TLM-2.0 documentation is as follows:

The default value of a transaction response status must be equal to "TLM_INCOMPLETE_RESPONSE."

Due to Definition 4.6, the formal definition of this statement is as follows:

$$(l_0 \; \&\& \; transact.tstatus \; == \; 0)) \; \rightarrow \; (l_1 \; \&\& \; transact.tstatus \; == \; 0) \qquad (4.6)$$
$$where \; \{l_0, \; l_1 \in E\}$$

As the temporal logic does not support enumerate types, the possible values of all TLM-2.0 enumeration types are denoted by integer values. In the case of the transaction response status attribute *transact.tstatus* the enumeration value *TLM_INCOMPLETE_RESPONSE* and *TLM_OK_RESPONSE* are defined by 0 and 1, respectively.

The TLM-2.0 *communication semantics* rules, on the other hand, define how communication must be carried out. The textbook specification of this rule in TLM-2.0 documentation is as follows:

A phase transition can only take place if the return-value of the non-blocking transport is TLM_UPDATED.

Based on Definition 4.6, the formal presentation of this rule is as follows:

$$(l_i \; \&\& \; tlm\_retun\_stus \; ! = \; 2 \; \&\& \; transact.tphase \; == \; 't') \; \rightarrow \qquad (4.7)$$
$$(l_{i+1} \; \&\& \; transact.tphase \; == \; 't') \; where \; \{l_i, l_{i+1} \in E\}, \; i > 0$$

The *tlm_retun_stus* is an integer variable that denotes the return value of the non-blocking transport function. The enumeration value *NO_RETURN, TLM_AC-CEPTED, TLM_UPDATED*, and *TLM_COMPLETED* are specified by 0, 1, 2, and 3, respectively. The *transact.tphase* parameter refers to the transition phase of the TLM transaction (such as *BEGIN_REQ*).

4.4.3.2   Translating VP's Specifications into Formal Properties

The design rules are usually written in textbook specifications and designers use them to implement the design. To model a SystemC VP, a part of these specifications is defined by the TLM-2.0 base protocol describing, e.g., how communication between TLM modules must be implemented. The other parts of these specifications, which are related to the functional and timing behavior of the VP, are defined by designers and considered as *User constraints*. In the following, we show how these constraints are translated into the corresponding proprieties.

The VP specifications, related to the functional constraints, are given as inputs based on the following definition.

Definition 4.7 The VP functional specifications VP_{fs}, for each initiator module *IM*, include the list of all target modules *TM* that *IM* is allowed to access them with a specific transaction type TT as below.

$$VP_{fs} = \{IM_i \mid IM_i \rightarrow \{(TM_j(address\_range),\ TT_n)\},$$

$$0 \le i \le n_{init}, 0 \le j \le n_{trg}, 0 \le n \le 13\}$$

where n_{init} and n_{trg} indicate the number of initiator and target modules, respectively.

The given VP specifications are analyzed by the *Property Generator* module (Fig. 4.1, phase 2) to aromatically generate all properties related to the invalid connection paths of an initiator and a target module either based on a wrong address or transaction types. As illustrated in Algorithm 4.1, the VP_{fs} is analyzed to extract all target modules of the VP and store them in the L_{trg} (line 1). For each initiator module *IM* of the VP, a *Forbidden Target List FTL$_{IM}$* is generated containing the target modules that the initiator is not allowed to access (lines 4 and 5). Then, properties are generated to indicate the following two important facts:

1. For the initiator module *IM*, there must be no connection path to the targets in its FTL_{IM} for all types of based protocol transactions specified in L_{TT} (lines 5–12).
2. For the initiator module *IM*, there must be no connection path to its target list *TL* with the transaction types specified in its *Forbidden Transaction Type* list FTT_{IM} (lines 13–20).

For example, consider the *LT_AT_BUS* VP (Fig. 3.1). The VP_{fs} of the design related to the *initiator_A* module is defined as below.

Algorithm 4.1: Functional Property (FP) generation

Data: Design specification VP_{fs}
Result: Functional properties FP to validate connection paths in SAP
1   $L_{trg} \leftarrow$ extracting all target modules of VP from VP_{fs};
2   $i \leftarrow 0$;
3   $FP \leftarrow \emptyset$;
4   **foreach** *target TM of initiator module IM in* VP_{fs} **do**
5      $FTL_{IM} \leftarrow L_{trg} - TM$;
6   **foreach** *target T in* FTL_{IM} **do**
7   **foreach** *transaction type TT in* L_{TT} **do**
8            $p_i \leftarrow (IM, T, TT)$;
9            $add(FP, p_i)$;
10            $i \leftarrow i + 1$;
11   **end**
12   **end**
13      $FTT_{IM} \leftarrow L_{TT} - TT_{IM}$;
14   **foreach** *target TM in TL* **do**
15   **foreach** *transaction type tt in* FTT_{IM} **do**
16            $p_i \leftarrow (IM, TM, tt)$;
17            $add(FP, p_i)$;
18            $i \leftarrow i + 1$;
19   **end**
20   **end**
21 **end**

$$VP_{fs} = \{Initiator\_A \rightarrow (Memory\_A(0x00 - 0x0A), T_1), \qquad (4.8)$$
$$(Memory\_A(0x0B - 0xFF), T_2),$$
$$(Memory\_B(0x00 - 0xFF), T_3)\}$$
$$(Memory\_C(0x00 - 0xFF), T_3)\}$$

Based on the aforementioned VP_{fs} and w.r.t Definition 4.6, a part of the functional properties FP, related to the communication path between *Initiator_A* and *Memory_A* $(0x00 - 0x0A)$, is generated as the following:

$$FP = \{p_i \mid p_i = (E <> \; !((l_0 \mid\mid l_1) \;\&\&\; (0x00 < address < 0x0A) \;\&\&\; TT_i))\}$$
$$(4.9)$$

where l_0 and l_1 denote the *initiator_A* and *Memory_A* modules, respectively. The parameter TT_i refers to the forbidden transaction types of the initiator module *initiator_A* to access the *Memory_A* module. The parameter *address* indicates the valid range of address that the initiator module *initiator_A* is allowed to access.

The property p_i in (4.9) indicates that there must be no access path to the address range $(0x00 - 0x0A)$ of the *Memory_A* (specified by location l_1) by the *initiator_A* (specified by location l_0) with the forbidden transaction type TT_i.

In order to validate the timing behavior of a given VP's transactions generated by different initiator modules against the VP's specifications, the timing specifications of the VP are required to be defined and given as inputs. These specifications are defined in the same way as the functional specifications. The only difference is that in addition to the information in Definition 4, the required time of a communication between an initiator module and its corresponding target (total transaction delay) needs to be identified in the VP's specifications. Thus, the VP_{ts} is defined as the following:

$$VP_{ts} = \{IM_i \mid IM_i \rightarrow \{(TM_j(address\_range),\ TT_n, TD)\},$$

$$0 \le i \le n_{init}, 0 \le j \le n_{trg}, 0 \le n \le 13\} \qquad (4.10)$$

where TD denotes the total delay of the generated transaction type TT by the initiator module IM to access the target module TM.

The *Property Generator* module (Fig. 4.1-phase 2) analyzes the given VP's specifications to generate timing properties automatically. As shown in Algorithm 4.2, for each target TM of the initiator IM, the property p_i indicates that there must not be a communication with the transaction type TT that requires less or more time than TD to perform the communication (lines 4–8). For example, a part of the generated timing properties TP related to the transactions generated by *Initiator_A* to access *Memory_A(0x00 − 0x0A)* is as the following:

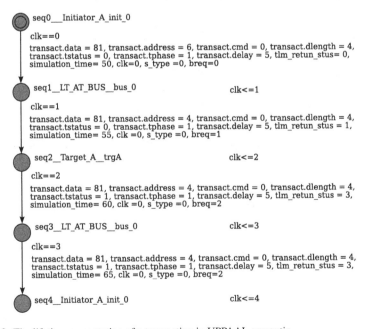

Fig. 4.3 The lifetime presentation of a transaction in UPPAAL semantic

---

Algorithm 4.2: Timing Property (TP) generation

Data: Design specification VP_{ts}
Result: Timing properties TP to validate connection paths in SAP
1   VP_{ts};
2   $i \leftarrow 0$;
3   $TP \leftarrow \emptyset$;
4   **foreach** *target TM of initiator module IM in* VP_{ts} **do**
5      $p_i \leftarrow (IM, TM, TT, !TD)$;
6      $add(TP, p_i)$;
7      $i \leftarrow i + 1$;
8   **end**

$$TP = \{p_1 \mid p_1 = E <> \; !((l_0 \mid\mid l_1) \;\&\&\; (0x00 < address < 0x0A) \;\&\&\; TT_1)$$
$$\&\& \; (TD! = 20))\} \tag{4.11}$$

where l_0 and l_1 denote the *initiator_A* and *Memory_A* modules, respectively. The parameter TT_1 refers to the transaction type with which the initiator module *initiator_A* is allowed to access the *Memory_A* module. The parameter *address* indicates the valid range of address that the initiator module *initiator_A* is allowed to access.

4.4.4 Compliance Checking

In order to verify the formal presentation of the simulation behavior of a given VP, UPPAAL [2]—a model checker that supports the timed automata model—is used. In UPPAAL, a query is a property that may or may not hold for the system. The UPPAAL query language is a subset of TCTL, denoting properties using a temporal logic language.

To verify the behavior of a transaction using UPPAAL, this behavior needs to be transformed into the UPPAAL semantics. To do this, the *UPPAAL Model Generator* module (Fig. 4.1, phase 3) performs the following steps:

1. Each transaction lifetime is defined as a system.
2. The queries are automatically generated from the template properties stored in the *Template properties* database.
3. The *UPPAAL Model* is generated, including both the timed automata model and queries.

Thus, the *Timed Automata Model* contains both the formal presentation of the simulation behavior of SystemC VP models (i.e., transaction lifetime and its access path representation) and the required properties to check the compliance of TLM-2.0 rules or the VPs' specifications.

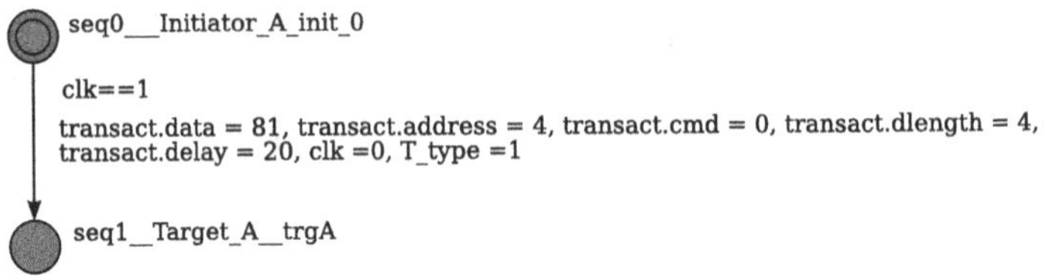

seq0___Initiator_A_init_0

clk==1
transact.data = 81, transact.address = 4, transact.cmd = 0, transact.dlength = 4,
transact.delay = 20, clk =0, T_type =1

seq1__Target_A__trgA

Fig. 4.4 The access path presentation of a transaction lifetime in UPPAAL semantic

The traced transactions of a VP are automatically defined as a system in UPPAAL semantics by the *UPPAAL Model Generator* module. To do this, the lifetime of each transaction is defined in terms of states (locations) and their transitions (edges). For each transaction lifetime (or its corresponding access path), the following steps are applied to map the transaction lifetime onto UPPAAL's timed automata concerning Definition 4.4 and 4.5 as below:

- The name of each system is denoted by *T_ID_number* showing a transaction with its ID and the number of its repetitions.
- Each sequence in the transaction lifetime is mapped onto a location (which is specified by the number of the sequence, the root name of the TLM module, and its instance name).
- The transaction's attributes and its related parameters of each sequence are mapped onto assignments for each transition.
- The invariant *clk* is defined to control the sequence of transitions.

As an example, Figs. 4.3 and 4.4 show the transaction lifetime and access path representation of a single transaction lifetime (illustrated in Fig. 4.2) of the *LT_AT_BUS* VP as UPPAAL systems, respectively. In both figures, the node with double circles specifies the initial state, and each node is specified by a name and an invariant. The assignment statements on each edge express the value of the transaction's attributes and all related parameters to describes its flow.

In Fig. 4.3, the transaction's related parameters are the simulation time *simulation_time*, the return value of function call *tlm_retun_stus*, the type of socket *s_type* and flags to check the timing phase *breq, ereq, brsp*, and *ersp*. The struct *transact* is defined to specify the value of transaction attributes (*data, address, cmd, dlength, tstatus*), the phase of the transition (*tphase*), and its timing annotations (*delay*) using integer values. As UPPAAL only supports integer values, all enumeration types of TLM-2.0 are mapped onto integer values. In Fig. 4.4, *Trans_type* indicates the transaction type.

The model checker gets the *UPPAAL model* (Fig. 4.1, phase 3) as an input file and verifies the system. The model checker reports the results of this compliance check, including the satisfied and violated properties

4.5 Experimental Evaluation

The proposed verification approach was applied to several standard VPs provided by Doulos [6] and [101]. The experimental results (Tables 4.1 and 4.2) are described in two parts. First, a real-world case study—the LEON3-based VP SoCRocket (implemented in SystemC TLM-2.0) [101]—is illustrated in detail in Sect. 4.5.1. Second, we give a brief discussion on the quality of the obtained experimental results in Sect. 4.5.2.

All the experiments have been carried out on a PC equipped with 8 GB RAM and an Intel Core i7 CPU running at 2.4 GHz.

4.5.1 Case Studies

To evaluate the quality of the proposed approach, we have injected faults into the VPs based on FT1, FT2, and FT3 introduced in Sect. 4.3. The proposed approach was applied to validate the correctness of each VP against the TLM-2.0 rules and the VP's specifications. These faults are injected into the VPs based on the aforementioned fault types as the following:

- FT1: an incorrect initialization of the transaction's response status (fault related to the transaction attributes rules), modification of the transaction data length by an interconnect module (fault related to the TLM modules behavior), and a wrong sequences order of transactions' phase transitions (fault related to the transaction's type).
- FT2: initiating transactions with an incorrect address computation or an incorrect initialization of the VP memory configuration file.
- FT3: altering the timing annotation of transactions with an incorrect computation.

For a real-world experiment, we used the proposed approach to validate the LEON3-based VP SoCRocket [101]. The VP is implemented in SystemC TLM-2.0 and includes more than 50,000 lines of code. It consists of several IPs working together in master (e.g., initiator modules *LEON3* processor, *ahbin1*, and *ahbin2*) or slave (e.g., target modules *AHBMem1* and *AHBMem2*) mode connecting to the on-chip bus *AHBCtrl* which is the AMBA-2.0 AHB (Advanced High-performance Bus). The communication uses a 32-bit address mode where the 12 most significant bits are used to specify the memory address. To show how different fault types are injected to the VP and the validation process was performed, consider a part of the VP including the initiator modules *ahbin1* and *ahbin2*, and the target modules *AHBMem1* and *AHBMem2* connected to the AMBA-2.0 AHB.

Regarding FT1, we injected the transaction attributes fault into the *ahbin2* to generate transactions with an incorrect default value of the response status attribute (i.e., TLM_OK_RESPONSE instead of TLM_INCOMPLETE_RESPONSE). We injected the TLM modules behavior fault into the AMBA-2.0 AHB to change the data length of receiving transactions (i.e., two instead of four bytes). Moreover, we

Table 4.1 Experimental results for all case studies related to the quality of the proposed verification approach

Variant	VP model	LoC	#Comp	#Trans	#TT	TM	#Queries			FTrans	R/VMU (MB)
							Total	Pass	Fail		
Original model	LT-example[1]	175	2	16	1	LT	128	128	0	0	6.7/41.5
	Routing-model[1]	456	6	10	1	LT	164	164	0	0	6.7/41.5
	Example-4[1]	547	2	10	4	AT	370	370	0	0	7.8/43.2
	Example-5[1]	650	7	10	1	LT	176	176	0	0	6.7/41.5
	Example-6[1]	713	9	20	2	AT	1332	1332	0	0	7.6/42.6
	AT-example[1]	3410	19	20	9	AT	1184	1184	0	0	7.8/43.2
	Locking-two[1]	4690	23	20	10	LT/AT	1420	1420	0	0	7.6/42.6
	SoCRocket[2]	50,000	20	200	8	LT/AT	9462	9462	0	0	7.8/43.2
FT1	LT-example[1]	175	2	16	1	LT	96	64	32	6	6.7/41.5
	Routing-model[1]	456	6	10	1	LT	120	112	8	4	6.7/41.5
	Example-4[1]	547	2	10	4	AT	350	332	18	3	7.8/43.2
	Example-5[1]	650	7	10	1	LT	120	110	10	5	6.7/41.5
	Example-6[1]	713	9	20	2	AT	1240	1145	95	9	7.6/42.6
	AT-example[1,a]	3410	19	20	9	AT	1060	924	142	7	7.8/43.2
	Locking-two[1,a]	4690	23	20	10	LT/AT	1280	1047	233	11	7.6/42.6
	SoCRocket[2]	50,000	20	200	8	LT/AT	8480	7806	674	37	7.8/43.2
FT2	LT-example[1]	175	2	16	1	LT	16	12	4	4	6.7/41.5
	Routing-model[1]	456	6	10	1	LT	24	18	6	6	6.7/41.5
	Example-4[1]	547	2	10	4	AT	10	8	2	2	7.8/43.2
	Example-5[1]	650	7	10	1	LT	40	30	10	10	6.7/41.5
	Example-6[1]	713	9	20	2	AT	60	45	15	10	7.6/42.6
	AT-example[1]	3410	19	20	9	AT	84	69	15	8	7.8/43.2
	Locking-two[1]	4690	23	20	10	LT/AT	92	75	17	9	7.6/42.6
	SoCRocket[2]	50,000	20	200	8	LT/AT	612	536	76	42	7.8/43.2

| Variant | VP model | LoC | #Comp | #Trans | #TT | TM | #Queries | | | FTrans | R/VMU (MB) |
							Total	Pass	Fail		
FT3	LT-example[1]	175	2	16	1	LT	16	10	6	6	6.7/41.5
	Routing-model[1]	456	6	10	1	LT	20	15	5	5	6.7/41.5
	Example-4[1]	547	2	10	4	AT	10	6	4	4	7.8/43.2
	Example-5[1]	650	7	10	1	LT	16	13	3	3	6.7/41.5
	Example-6[1]	713	9	20	2	AT	32	28	4	4	7.6/42.6
	AT-example[1]	3410	19	20	9	AT	40	30	10	10	6.7/41.5
	Locking-two[1]	4690	23	20	10	LT/AT	48	42	6	6	7.8/43.2
	SoCRocket[2]	50,000	20	200	8	LT/AT	370	311	59	59	7.8/43.2

LoC lines of code, *#Trans* number of transactions, *#TT* number of transactions' types, *TM* timing model, *FTrans* number of faulty transactions, *R/VMU* resident/virtual memory usage peaks

[1]Provided by Aynsley [6]

[2]Provided by Schuster et al. [101]

[a]The *AT-example* and *Locking-two* VPs include custom base protocol checkers in their original source codes and were removed to create faulty model FT1

Table 4.2 Experimental results for all case studies related to the execution time of the proposed verification approach

Variant	VP model	LoC	#Comp	#Trans	#TT	TM	Phase1 (s)		Phase2 (s)	Phase3 (s)	Total (s)	
							CbA	DbA			CbA	DbA
Original mode	LT-example[1]	175	2	16	1	LT	1.21	8.06	0.16	0.15	1.52	8.37
	Routing-model[1]	456		10	1	LT	2.18	21.49	0.26	0.29	2.73	22.04
	Example-4[1]	547	2	10	4	AT	2.32	86.53	0.32	2.38	5.02	89.23
	Example-5[1]	650	7	10	1	LT	3.29	92.65	0.28	1.06	4.63	93.99
	Example-6[1]	713	9	20	2	AT	5.04	137.39	0.82	3.64	9.50	141.85
	AT-example[1]	3410	19	20	9	AT	19.11	469.32	0.73	3.17	23.02	473.23
	Locking-two[1]	4690	23	20	10	LT/AT	25.84	611.48	0.89	3.73	30.46	616.10
	SoCRocket[2]	50,000	20	200	8	LT/AT	54.79	5491.05	3.34	5.02	63.15	5498.41
FT1	LT-example[1]	175	2	16	1	LT	1.21	8.06	0.16	0.15	1.52	8.37
	Routing-model[1]	456	6	10	1	LT	2.18	21.49	0.26	0.29	2.73	22.04
	Example-4[1]	547	2	10	4	AT	2.32	86.53	0.32	2.38	5.02	89.23
	Example-5[1]	650	7	10	1	LT	3.29	92.65	0.28	1.06	4.63	93.99
	Example-6[1]	713	9	20	2	AT	5.04	137.39	0.82	3.64	9.50	141.85
	AT-example[1,a]	3410	19	20	9	AT	19.11	469.32	0.73	3.17	23.02	473.23
	Locking-two[1,a]	4690	23	20	10	LT/AT	25.84	611.48	0.89	3.73	30.46	616.10
	SoCRocket[2]	50,000	20	200	8	LT/AT	54.79	5491.05	3.34	5.02	63.15	5498.41
FT2	LT-example[1]	175	2	16	1	LT	1.21	8.06	0.04	0.05	1.20	8.15
	Routing-model[1]	456	6	10	1	LT	2.18	21.49	0.05	0.05	2.28	21.59
	Example-4[1]	547	2	10	4	AT	2.32	86.53	0.02	0.03	2.37	86.58
	Example-5[1]	650	7	10	1	LT	3.29	92.65	0.06	0.05	3.40	92.76
	Example-6[1]	713	9	20	2	AT	5.04	137.39	0.07	0.05	5.16	137.51
	AT-example[1]	3410	19	20	9	AT	19.11	469.32	0.08	0.06	19.25	469.46
	Locking-two[1]	4690	23	20	10	LT/AT	25.84	611.48	0.08	0.06	25.98	611.62
	SoCRocket[2]	50,000	20	200	8	LT/AT	54.79	5491.05	0.15	0.08	55.02	5491.28

Variant	VP model	LoC	#Comp	#Trans	#TT	TM	Phase1 (s)		Phase2 (s)	Phase3 (s)	Total (s)	
							CbA	DbA			CbA	DbA
FT3	LT-example[1]	175	2	16	1	LT	1.21	8.06	0.02	0.04	1.17	8.12
	Routing-model[1]	456	6	10	1	LT	2.18	21.49	0.02	0.03	2.26	21.55
	Example-4[1]	547	2	10	4	AT	2.32	86.53	0.01	0.02	2.35	86.56
	Example-5[1]	650	7	10	1	LT	3.29	92.65	0.04	0.04	3.37	92.73
	Example-6[1]	713	9	20	2	AT	5.04	137.39	0.05	0.04	5.13	137.48
	AT-example[1]	3410	19	20	9	AT	19.11	469.32	0.05	0.04	19.20	469.41
	Locking-two[1]	4690	23	20	10	LT/AT	25.84	611.48	0.06	0.05	25.95	611.59
	SoCRocket[2]	50,000	20	200	8	LT/AT	54.79	5491.05	0.09	0.06	54.95	5491.21

CbA compiler-based approach, *DbA* debugger-based approach, *Phase1* the required time for information extraction and translation, *Phase2* the required time for property generation, *Phase3* the required time for UPPAAL model generation and validation

[a]The *AT-example* and *Locking-two* VPs include custom base protocol checkers in their original source codes and were removed to create faulty model FT1

[1]Provided by Aynsley [6]

[2]Provided by Schuster et al. [101]

injected the transaction type fault into the *ahbin1* to generate transactions with initial phase BEGIN_RESP instead of BEGIN_REQ for transaction type T_1.

The first fault was detected by checking the queries related to the transaction attribute rules in the generated UPPAAL models. For example, the following query is automatically generated from the template properties for the transaction lifetime *TL101*.

$$(TL101.seq0\_ahbin2\_init2 \ \&\& \ transact.tstatus \ == \ 0)) \rightarrow \qquad (4.12)$$

$$(TL101.seq1\_AHBCtrl\_ahb \ \&\& \ transact.tstatus \ == \ 0)$$

This query indicates that the initial response statues in the 101st transaction lifetime, generated by the initiator module *ahbin2* to access the target module *AHBMem2* through *AHBCtrl* interconnect, must be TLM_INCOMPLETE_RESPONSE.

The proposed approach was able to find the second fault by checking the queries related to the TLM module behavior. For example, the following query is automatically generated from the template properties for the transaction lifetime *TL369*.

$$(TL369.seq1\_AHBCtrl\_ahb \ \&\& \ transact.dlength \ == \ 4)) \rightarrow \qquad (4.13)$$

$$(TL369.seq2\_AHBMem2\_trg2 \ \&\& \ transact.dlength \ == \ 4)$$

This query checks the interconnect module *AHBCtrl* does not modify the data length attribute of transactions.

The third fault was detected by checking the queries related to the transaction type. For example, the following query is automatically generated from the template properties for the transaction lifetime *TL130*.

$$E <> \ ((TL130.seq0\_ahbin1\_init1 \ || \ TL130.seq1\_AHBCtrl\_ahb) \qquad (4.14)$$

$$\&\& \ (transact.tphase \ == \ 1))$$

The above query indicates that the phase transition between the first two sequences of a transaction lifetime must be *BEGIN_REQ*. With this regard, the 130th generated transaction of the initiator module *ahbin1* must have the transition phase (*tphase*) *BEGIN_REQ* (specified by 1) when received by the interconnect module *AHBCtrl*.

Overall, 30 transactions were generated by both initiator modules, which 11 of them were against the TLM 2.0 rules.

Concerning FT2, we consider the expected functional specifications of the VP as the following:

$$VP_{fs} = \{ahbin1 \rightarrow (AHBMem1(0xA0000000 - 0xA0000CC4), T_1), \qquad (4.15)$$

$$(AHBMem1(0xA0000CC5 - 0xA0000FFF), T_2),$$

$$(AHBMem2(0xB0000000 - 0xB0000FFF), T_8),$$

$$ahbin2 \rightarrow AHBMem2(0xB0000000 - 0xB0000FFF), T_0)\}$$

The functional faults are injected into the lines of code of the *ahbin1* where transactions address are generated. These lines of code were altered to generate random values for the 12 least significant bits of the transactions' address (i.e., *000* to *FFF*) for both transaction types T_1 and T_2.

Due to the VP_{fs}, 80 queries were automatically generated from the corresponding eight template properties in the template functional properties of the VP by the *UPPAAL Model Generator* module (Fig. 4.1, phase 3). For example, the following query indicates that in the access path *AP54*, there must be no path with the transaction of types T_0, T_2, T_3, ..., T_{13} generated by *ahbin1* to access the address range ($0xA0000000- 0xA0000CC4$) of the *AHBMem1* module.

$$E <> !((AP54.seq0\_ahbin1\_init1 \;||\; AP54.seq1\_AHBMem1\_trg1) \qquad (4.16)$$
$$\&\& \; (T\_type == 0 \;||\; T\_type == 2 \;...\; ||\; T\_type == 13)$$
$$\&\& \; (transact.address > 0 \;\&\&\; transact.address < 3268))$$

Overall, six queries were violated, indicating that the initiator modules *ahbin1* and *ahbin2* had six invalid access to the target modules *AHBMem1* and *AHBMem2* w.r.t the functional specifications of the VP. The violated queries are related to four transactions out of 30 generated transactions by the initiator modules *ahbin1* and *ahbin2*.

Regarding FT3, we consider the expected timing specifications of the VP related to the *ahbin1* and *ahbin2* as below.

$$VP_{ts} = \{ahbin1 \rightarrow (AHBMem1(0xA0000000 - 0xA0000CC4), T_1, 50), \qquad (4.17)$$
$$(AHBMem1(0xA0000CC5 - 0xA0000FFF), T_2, 100),$$
$$(AHBMem2(0xB0000000 - 0xB0000FFF), T_8, 200),$$
$$ahbin2 \rightarrow (AHBMem2(0xB0000000 - 0xB0000FFF), T_0, 50)\}$$

We changed the lines of code where the timing annotation of the transactions type T_1 and T_2 generated by *ahbin1* are defined. Due to the VP_{ts} in (4.17), four timing queries were generated. For example, the following query checks that the access path *AP54* adheres the VP timing specifications related to the *ahbin1* module.

$$E <> !((AP54.seq0\_ahbin1\_init1 \;||\; AP54.seq1\_AHBMem1\_trg1) \qquad (4.18)$$
$$\&\& \; (T\_type == 1) \;\&\&\; (transact.delay \,!= 50)$$
$$\&\& \; (transact.address > 0 \;\&\&\; transact.address < 684357828))$$

Our validation approach could detect two properties violation. From 30 generated transactions by the initiator modules *ahbin1* and *ahbin1*, three transactions had incorrect timing behavior w.r.t the VP_{ts}.

The experimental results for different types of ESL benchmarks are shown in Tables 4.1 and 4.2. In both tables, the first column shows four variants of SystemC VPs denoted as *Original*, *FT1*, *FT2*, and *FT3* referring to the reference model of the VP and three faulty models, respectively. Columns *VP Model, Loc*, and *#Trans* list name, lines of code, and the number of extracted transactions for each VP, respectively. The *#TT* column illustrates the number of transactions type implemented in each VP. Column *TM* presents the timing model of each design.

In Table 4.1, Column *#Queries* shows the number of generated queries (properties) to validate each VP against the TLM-2.0 rules or its specifications. For this column, *Total, Pass, Fail*, and *FTrans* illustrate the number of generated, satisfied and violated queries, and the number of faulty transactions, respectively. For the original model of each VP, column *properties* show the total number of all generated queries that the VP was validated against them. The value zero for columns *Fail* and *FTrans* indicate that all transactions adhered the TLM-2.0 rules and the VP specifications. The *R/VMU* column shows the value of resident memory over virtual memory usage peaks when checking a property using the UPPAAL model checker. As UPPAAL considers each query separately, for each system, the maximum R/VMU was reported.

The execution time of the proposed approach is reported in Table 4.2. In this table, column *Phase1* shows the required time for information extraction and translation into the prospered structured followed by the data extraction using the compiler-based approach (*CbA*) and debugger-based approach (*DbA*). Column *Phase2* presents the time that the *property generation* module (Fig. 4.1, phase2) requires to generate the corresponding properties for all transactions' lifetime or access paths. Column *Phase3* illustrates the required time to generate UPPAAL models and their validation. Column *Total* shows the total execution time of the proposed verification approach. Despite from the first phase execution time of the proposed approach (which is the major time-consuming part and was discussed in detail in Chap. 3, Sect. 3.6), the execution time of other phases is mostly less than a second. Overall, the total execution time using the compiler-based approach even for a complex VP (SoCRocket) is about one minute, providing a fast verification solution. In the case of debugger-based approach, the total execution time is still within a reasonable time frame.

4.5.2   Integration and Discussion

The proposed approach is able to automatically retrieve the simulation behavior of a given SystemC TLM-2.0 VP and formally verify it against both the TLM-2.0 rules and the VP's specifications.

In comparison to the formal verification methods that are restricted to a subset of SystemC models because of several pre-conditions and pre-assumptions on SystemC implementations, the proposed method has no limitation on how and which SystemC constructs are used to implement an ESL design. Although the proposed approach does not assure the validity of the model like formal methods (as it verifies the model's simulation behavior instead of its formal specification), it makes a trade-off between the accuracy of the model to check the compliance of the model and its applicability to verify a wide range of SystemC designs.

Unlike simulation-based verification approaches that rely on modifying the SystemC library, interfaces or kernel to verify the run-time information, the proposed approach verifies the detailed simulation behavior of a given TLM model without any modification to the aforementioned standard tool flow. It means that the proposed approach provides designers with an easy-to-use and automated verification solution.

4.5.2.1 Limitations

As the proposed approach is based on run-time analysis, it inherits the same limitations. To properly verify a given VP, the running application of benchmark must activate different communication flow and transactions implemented in the VP. However, this can be done either using manual or automated test generation techniques.

4.6 Conclusion

In this chapter, a comprehensive verification approach was presented, enabling designers to validate a given SystemC-based VP implemented using SystemC TLM-2.0 framework against the TLM-2.0 rules, and the VP's specifications. The approach is based on formally verifying the simulation behavior of VPs by UPPAAL model checker. The simulation behavior, which describes the transactions VPs, is extracted automatically using the suggested SystemC VP analysis approaches in Chap. 3. The extracted transactions are defined formally in terms of timed automata models and a set of access paths. A formal description of a large set of the TLM-2.0 rules and the VP specifications (related to the functionality and timing behavior of IPs' communication in VPs) are provided into a set of template properties in the language of temporal logic TCTL. Concerning the TLM-2.0 rules, three types of potential violations related to the transaction attributes, communication flow, and TLM modules behavior were defined and injected into a set of extensive case studies to evaluate the proposed approach. Regarding the validation of VPs against their specifications, VPs were validated against two types of fault related to the functionality and timing behavior of IPs' communication. The experimental results confirmed the applicability and quality of the proposed approach to verify various TLM-2.0 VPs with different complexity.

Chapter 5
Application II: Security Validation

Modern SoCs are notoriously insecure. Hence, the fundamental security feature of IP isolation is heavily used, e.g., secured *Memory Mapped IOs* (MMIOs), or secured address ranges in case of memories, are marked as non-accessible. One way to provide reliable assurance of security is to define isolation as information flow policy in hardware using the notion of non-interference. Since insecure hardware opens up the door for attacks across the entire system stack (from software down to hardware), the security validation process should start as early as possible in the SoC design cycle, i.e., at the ESL.

In this chapter, we show another application of the design understanding in the design process for the task of security validation at the ESL. We take advantage of the introduced VP analysis approaches in Chap. 3 to perform a dynamic information flow tracking analysis to access the run-time behavior of a given VP-based SoC. The extracted behavior is validated against security threat models, such as information leakage (confidentiality) and unauthorized access to data in a memory (integrity).

5.1 Introduction

Transistor by transistor, the electronics industry is changing the world. With billions of people connected by nearly a trillion "secure" electronic devices, the evolution of the SoC design process in the electronics industry was inevitable. To meet the increasing demand of SoCs, the modern SoC design process has shifted from "*in-house*" development of IPs to "*use and reuse*" of existing commercial IPs. Hence, modern SoCs consist of millions of lines of code and often leverage libraries, tool-kits, and components from third-party vendors. This paradigm shift has significantly lowered the design costs and reduced the time-to-market. As a consequence of this decentralization, modern SoCs are notoriously insecure where third-party IPs, in particular, can be used as a vehicle for malice. For example, the recent security

© Springer Nature Switzerland AG 2020
M. Goli, R. Drechsler, *Automated Analysis of Virtual Prototypes at the Electronic System Level*, https://doi.org/10.1007/978-3-030-44282-8_5

compromise of SoCs using Intel's microprocessor IPs, and Actel ProASIC3 IPs. The former were exploited by *"Meltdown and Spectre"* vulnerabilities [64, 71], and the latter with JTAG vulnerability [102]. This highlights the fact that legitimate commercial off-the-shelf IPs may manipulate or assist in manipulating secure user information in ways that their users neither expect, nor appreciate. Hence, the fundamental security feature of IP isolation (or memory isolation) is considered, e.g., secured address ranges or IPs are marked as non-accessible.

While the existing validation methodologies are effective in securing either the software or the hardware IP individually, holistic system-level solutions targeting the entire SoC (particularly composed of third-party software programs and hardware IPs) are lacking. The software security validation assumes trustworthy hardware, and the hardware IP security validation assumes secure software (firmware/Operating System) running on top. As mentioned earlier, one way to provide reliable assurance of security for the entire SoC is to define IP isolation as information flow policy in hardware using the notion of non-interference. This can close the semantic gap across the entire system stack (from software down to hardware), and protect against malicious software which may exploit hardware backdoors to cause malfunctions or leak confidential information. Hence, instead of considering security validation as an individual process starting at a later stage, the *Secure Development Lifecycle* [62] practices should be adopted from the start of the SoC design process (i.e., at the ESL). VP-based security validation could be one promising direction to fix the security vulnerabilities in the SoCs before they are refined and expensive design loops occur.

One possible solution for security validation is *Information Flow Tracking* (IFT), which can detect security vulnerabilities that violate certain information flow security policies and properties (commonly non-interference properties of confidentiality, and integrity). IFT associates each data object with a label and monitors the flow of labeled information through the entire system. IFT techniques (static and dynamic) have been proved to be effective in enforcing security policies at various levels of the system stack, i.e., from compiler to gate-level design. IFT can be performed at the hardware level [4, 55, 108, 114], or in software through the use of source-level instrumentation [32, 119], or binary instrumentation [19, 61, 90, 115]. There has not been much work on system-level VPs [50], in particular on binary VP models which are common when integrating third-party IPs, and hence, the source code is not available.

In this work, we propose a VP-based security validation approach using a combination of dynamic IFT and post-execution analysis. To perform IFT (tracing TLM transactions), we take advantage of the information extraction approaches introduced in Chap. 3. The post-execution analysis includes transactions translation, security property generation, and finally, VP model validation against security properties, to identify the TLM transactions, and the dynamic paths they take to violate the security properties. The focus of the proposed approach is to detect the security violations related to the most occurring threat models, which are confidentiality and integrity. The precise violation paths are reported back to the verification engineer to either replace the (third party) VP model, or update the security policy.

The rest of this chapter is organized as follows. The literature review is presented in Sect. 5.2. Then, the threat models with a motivating example are presented in Sect. 5.3. The proposed methodology for dynamic information flow tracking, security properties generation, and validation is introduced and explained in Sect. 5.4. Experiments on several VPs to show the efficacy of our approach are presented in Sect. 5.5. Finally, the chapter is concluded in Sect. 5.6.

5.2 Related Works

Software IFT like LIFT [90], Minemu [11], and RIFLE [115] use dynamic binary translation to detect security vulnerabilities. These methods are fast and optimized, but they are focused on single problem domains. Extending them to different architectures requires extensive modifications.

On the other hand, researchers have proposed another approach for hardware trustworthiness; Proof-Carrying Hardware [22, 46, 57, 73], inspired by Proof-Carrying Code [21, 22]. It verifies the equivalence between design specification and design implementation using a run-time Combinational Equivalence Checking, which makes it very effective. However, it does not consider security properties verification.

Gate Level IFT (GLIFT) [114] is another novel approach used at design time for testing and verification. Because of gate-level abstraction, GLIFT is not limited to micro-architectural units only, and rather it can be applied to any digital hardware system. It enables GLIFT to track each bit in the system, irrespective of the security properties. To do so, it generates a separate analysis logic (shadow logic) for each gate for IFT analysis, which tracks a different set of flows by labeling hardware variables. This extra logic can be fabricated and used at run-time or instantiated in the design phase to verify whether the hardware design enforces the desired information flow policies. However, under certain scenarios, the generated logic is over-approximated, i.e., false positives happen. GLIFT covers both implicit flows, and timing channels, as the gate-level makes all information flows explicit. While GLIFT is able to identify any information flow violations precisely, it requires a gate-level design.

On the other hand, RTLIFT [4] follows GLIFT but raises the abstraction level to RTL. It gives the flexibility to define both implicit and explicit flows. It encodes security attributes into the design for formal verification of hardware security properties. However, the approach is limited to a single IP core.

Caisson [69], VeriCoq-IFT [9], Sapper [70], and SecVerilog [120] are hardware security design languages. While there are many advantages of the aforementioned language level works, the major disadvantages are new language familiarity, overestimation of the flows, and re-labeling of designs each time a new property is added among others.

Recently, researchers have proposed a VP-based IFT solution [50] which works on SystemC TLM-2.0 abstraction and statically analyzes the complete SoC for information flow. The approach uses Clang library to perform binding analysis, access control extraction, call graph analysis, data flow analysis [49], and static taint analysis to determine if the specified security properties have been satisfied or not. The violated properties, along with potential violation paths, are reported back to the user. The approach identifies all the potential vulnerable paths as an over-approximation, i.e., no path is missed. However, the approach suffers from the following shortcomings: over-approximation of flows leading to false positives, inability to precisely identify a path in case of dynamic variables.

Our proposed approach eliminates the drawbacks of the aforementioned approaches. It works directly on the VP models. The verification engineer does not require to learn new hardware security language, the analysis is dynamic at run-time, and the security properties are also defined only once.

5.3   Motivation and Threat Models

In this section, we introduce the threat models considered in our work and show a motivating example to demonstrate how the threat models affect the security of a VP-based SoC.

5.3.1   Threat Models

To show the importance and criticality of security validation in a given SoC design, it is necessary to identify (1) which types of assets must be protected and (2) what kind of threats we are protecting against. Assets such as sensitive configuration registers, cryptography keys, and digital signatures are the critical and confidential information that must be protected against unauthorized access. According to [27, 50] the threat models of leaking information in a given SoC design at the ESL can be divided into two main categories:

- **Confidentiality:** an unauthorized IP retrieves data of secure IP (e.g., data in a secure memory).
- **Integrity:** an unauthorized IP modifies data of secure IP.

For a given SoC design, two common potential sources of the threat models mentioned above are as the following:

1. The design contains no (strong) security policies (i.e., access control or information flow policy). This makes the design exploitable by vulnerabilities. For example, when an SoC includes a hardware IP purchased from an untrusted third-party vendor, the IP can contain malicious part to leak the confidential data.

In the same way, an incorrect initialization (either by an adversary involved in the IP design process or unintentionally) of the SoC firmware (e.g., memory configuration file) can cause an unauthorized IP to access the sensitive data in the secure memory. In both cases, well-implemented security policy in the SoC could prevent sensitive data from being leaked.

2. The existing SoC is extended/modified, but its security policy is not updated. Especially in case that the design team decides to improve the existing SoC by adding new IPs (e.g., an accelerator, a processor, or a memory), the previous security policies may not be sufficient to protect sensitive data against leakage.

Therefore, for a given SoC, the security validation analysis must be performed to ensure that the secure assets cannot be inferred either with *direct access* (direct communication of two IPs) or through *indirect access* (IP collusion).

5.3.2 Motivating Example

Consider the *RISC-32 SoC* model shown in Fig. 5.1. The VP is inspired by the example (Fig. 3.1) in Chap. 3.

The hardware part consists of two initiator modules: *RISC-CPU* and *Initiator_0*, a generic loosely timed interconnect *LT_BUS* and two memories *M0_regular* (regular memory) and *M2_secure* (secure memory). The interconnect module is accompanied with a specifications sheet. It can support up to four initiators and eight target modules based on the following configuration: (1) the target modules connected to the *LT_BUS* initiator ports 1 and 8 are only accessible by initiator modules connected to the *LT_BUS* target ports 1 and 4, respectively, and (2) the target modules connected to the *LT_BUS* initiator port 2 to 7 are shared with all four initiators. *M2_secure* stores sensitive data (connected to initiator port 1 of the *LT_BUS*), and *M0_regular* (connected to initiator port 8 of the *LT_BUS*) is used to store normal computation results. The *RISC-CPU* module (connected to target port 1 of the *LT_BUS*) is a standard commercial 32-bit RISC-CPU provided by Accellera Systems Initiative [1] and extended to support TLM-2.0 and C++ as its running

Fig. 5.1 The architecture of the motivating example *RISC-32 SoC* VP

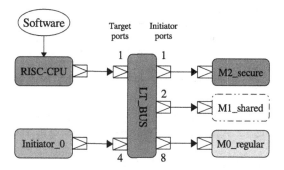

software. The initiator *Initiator_0* is a processing element which is connected to target port 4 of the *LT_BUS*. Memory *M1_shared* (connected to initiator port 2 of the *LT_BUS*) is not initially available.

The software part (running on the *RISC-CPU*) includes a key generation routine for an encryption algorithm (to simplify, the actual functionality is abstracted away), a compiler (to translate the software into *RISC-CPU* instructions), and a memory configuration file (used by the compiler for memory allocation). The memory configuration file for the *RISC-CPU* initially includes the information that the only available memory is *M2_secure*. To reduce the risk of run-time errors due to memory management, memory is allocated at compile-time which is mostly used in SoC designs. Figure 5.2 demonstrates a part of the running software *foo_64* (lines 1 to 10) of the *RISC-CPU* generating a 64-bit encryption key. Variables *key*1 (line 3) and *key*2 (line 4) are initialized by *init*1 and *init*2 which are extracted from the *init_key* (line 1) file. The final key *final_key* (line 7) is generated after some intermediate computation by *key_gen* function. As *M2_secure* is the only available memory for the *RISC-CPU* in the memory configuration file, all variables are mapped by the compiler into this memory.

According to the aforementioned configuration w.r.t the interconnect specification, the following security policies are satisfied for the *RISC-32 SoC* design:

- *M0_regular* is only accessible by *initiator_0* and
- *M2_secure* is only accessible by *RISC-CPU*.

```
1     void foo_64 (fstream &init_key){
2     /* ... */
3     unsigned char key1[8] = init1;
4     unsigned char key2[8] = init2;
5     unsigned char final_key[8] = {0};
6     /* ... */
7     final_key = key_gen (key1,key2);
8     /* ... */
9     return;
10    }
11    /*The function is upgraded to generate more robust key*/
12    void foo_256 (fstream &init_key){
13    /* ... */
14    unsigned char key1[32] = init1;
15    unsigned char key2[32] = init2;
16    unsigned char key3[32] = init3;
17    unsigned char final_key[32];
18    /* ... */
19    final_key = key_gen (key1,key2,key3);
20    /* ... */
21    return;}
```

Fig. 5.2 A part of *RISC-CPU* running software of the motivating example

Now consider the scenario that the software is upgraded (*foo_256*—line 12 in Fig. 5.2) to generate a more robust key by increasing its length (from 64-bits to 256-bits) and modifying the *key_gen* (line 19 in Fig. 5.2) function by adding one more initial key *key3* to its input arguments. After compiling the new software, an error is generated by the compiler stating that the memory is not sufficient. To overcome this problem and to increase the overall performance of the design, memory *M1_shared* is added to the *RISC-32 SoC*. The memory configuration file needs to be updated, allowing the compiler to use the new memory space to map data. Thus, it might be possible that some variables (such as *final_key*—line 19 in Fig. 5.2) containing the secure data (that have been already located in *M2_secure* before adding the new memory) are now mapped into the *M1_shared*. Therefore, at run-time, an unwanted information flow can cause sensitive data to be read from the secure memory and then written to the shared memory. The main reason for such vulnerability is that the interconnect security policy is not updated after inserting a new IP (in the example at hand *M1_shared*). Thus, the underlying hardware can be used as a gateway by an unprivileged software to leak the confidential data. Since the security policy of the *LT_BUS* has not been updated, the running software (i.e., application) can use the *RISC-CPU* to read the sensitive data from *M2_secure* and write it to *M1_shared*. Subsequently, *initiator_0* can read the secure data from *M1_shared*.

Detecting this type of indirect access even for a simple SoC model is not a trivial task as it cannot be detected either by functional verification methods (as the functionality of the SoC model is still correct) or using static security validation analysis. In case that the address of transactions is defined at run-time, e.g., generated either explicitly by initiator modules (based on some dynamic computation) or implicitly by its running software (like in the example above), static analysis approaches are not able to detect this security violation. Hence, a dynamic information flow analysis is required such that the run-time behavior of a given SoC is verified against the threat models in both *direct access* and *indirect access* cases.

5.4 Methodology

In this section, we first give an overview of the proposed security validation approach. Then, four phases of the approach are explained in detail.

5.4.1 Overall Workflow

Figure 5.3 provides an overview of the proposed approach which consists of four main phases as the following:

1. Tracking transactions of a given VP model at run-time.
2. Translating the transactions into set of access paths.

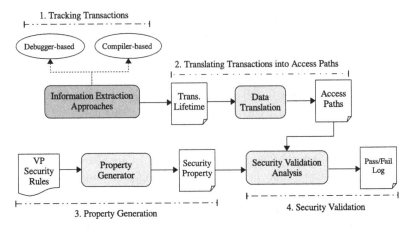

Fig. 5.3 The proposed security validation methodology overview

3. Generating security properties from the design's security rules.
4. Validating the translated model of the design against the security properties.

The information related to each transaction is extracted only once during the execution time and translated into a set of access paths. Each path identifies an information flow from the *source* (the creation point of a transaction by an initiator IP) and *destination* (the final point of the transaction which is in a target IP). The design security rules are translated into set of security properties based on the direct and indirect access scenarios. In the final phase, the security validation analysis ensures that there is no path where an unauthorized *source* is connected to a protected *destination*. The approach identifies the failing paths for each unsatisfied security property, allowing the design team to focus on the exact vulnerable points of the design and improve it either by enhancing access control policies or information flow policies [43]. In the following, each phase of the proposed approach is explained in detail and illustrated using the motivating example of Sect. 5.3.

5.4.2   Tracking Transactions

IPs in an SoC at VP level communicate through TLM transactions. To validate that no information leaks, a complete trace of all transactions is required. Especially, the *indirect access* can only be detected when the whole behavior of the design, including all intermediate communication of (initiator and target) IPs, is analyzed.

To properly track a given VP's transactions, we take advantage of utilizing the information extraction approaches presented in Chap. 3 to access the run-time information of a given TLM design. As the target of the security validation analysis is to find whether or not the initiator IPs (source) and the target IPs (destination) communicate based on the VP's security rules, we only need to extract the run-time information related to a transaction at these two points. Hence, for this task, we use the filter options of the information extraction approaches presented in Chap. 3, Sect. 3.4.3 to extract only the run-time information of transactions that is generated by initiator modules and received by the corresponding target modules. Thus, the run-time information of transactions is not extracted when they pass through interconnect IPs. Moreover, for each transaction we extract only the attributes, e.g., data, address and data length, and its related parameters such as reference address of transaction object, transition phase, and return value of function call, that are required for our security validation analysis. This speeds up our information flow analysis as the mount of data, that needs to be extracted, is reduced.

5.4.3   Translating Transactions into Access Paths

As mentioned earlier in Chap. 4, the VP-based SoC behavior can be seen in terms of abstract communication paths illustrating how transactions flow through the system (from *source* to *destination*). However, to validate a given VP-based SoC against its security rules, the definition of access path presented in Definition 4.3 needs to be modified. As the security rules of a given design are defined based on communication between the initiator and the target modules, each access path must be defined based on the information related to the *source* and *destination* of the corresponding transaction as below.

Definition 5.1

$$AP = \{P_i \mid P_i = \{TID \rightarrow IM \rightarrow cmd \rightarrow TM \rightarrow$$
$$adrs \rightarrow L \rightarrow ST\} ; \ 1 \leq i \leq n_T\}$$

where

- *TID* is the transaction ID including transaction reference address (to distinguish the generated transactions of different initiator modules),
- *IM/TM* is initiator/target module which consists of its root name, the instance name, and the name of its function call (it defines the start/end point of the transaction),
- *cmd* is the command attribute of the transaction,
- *adrs* is the address attribute of the transaction (indicating transaction address in TM),
- *L* is the *data_length* attribute of the transaction,
- *ST* is the simulation time stamp of transactions, and
- n_T is the number of transactions.

The *Trans Lifetime* is analyzed by the *Data Translation* module (Fig. 5.3, phase 2) to extract all access paths embedded in the transactions' lifetime and present them based on Definition 5.1. Each member of the *AP* set indicates a

```
AP_RISC32−SoC = {

P1=(0x6761b0_0 -> initiator_0 :: init_0 :: thread_process -> READ ->
M0_regular :: mem0 :: b_transport -> 0x004 -> 4-> 0ns),

P2=(0x683c10_0 -> RISC−CPU :: init_1 :: thread_process -> READ ->
M2_secure :: mem2 :: b_transport -> 0x010 -> 4-> 100ns),

P3=(0x6761b0_1 -> initiator_0 :: initi_0 :: thread_process -> WRITE
  ->
M0_regular :: mem0 :: b_transport -> 0x008 -> 4-> 200ns),

P4=(0x683c10_1 -> RISC−CPU :: init_1 :: thread_process -> WRITE ->
M1_shared :: mem1 :: b_transport -> 0x004 -> 4-> 300ns),

P5=(0x6761b0_2 -> initiator_0 :: init_0 :: thread_process -> READ ->
M1_shared :: mem1 :: b_transport -> 0x004 -> 4-> 400ns)}
```

Fig. 5.4 A part of the generated access paths of the *RISC-32-SoC* VP

direct flow of information between *IM* and *TM*, thus, declared as a direct path.
A combination of three direct paths specifies an indirect information flow if it
enables an initiator to access the data of a target which is inaccessible directly. For
instance, Fig. 5.4 illustrates a part of the generated *RISC-32-SoC* design's access
paths. In this set, *P1*, *P2*, ...and *P5* present five direct paths. Paths *P1*, *P3*, and
P5 indicate three transactions generated by *initiator_0* to access data of regular
and shared memories *M0_regular* and *M1_shared*, respectively. For indirect path,
consider the combination of paths *P2*, *P4*, and *P5* creating an indirect data flow
between *initiator_0* and *M2_secure*. It demonstrates that in path *P2* the *RISC_CPU*
generates a transaction to read data from secure memory *M2_secure*. In path *P4*,
RISC_CPU writes the read secure data into shared memory *M1_shared* and finally
in path *P5* the secure data is read by *initiator_0*. Thus, *initiator_0* could access the
secure data using *RISC_CPU*.

5.4.4 Security Property Generation

The design security rules need to be expressed in a well-defined format in order to
validate the SoC. The security rules are usually written in a textbook specification
(defined as reference model) where designers use it to implement the access control
policies of the interconnect module or other mechanisms to prevent information
leakage. The security rules of the SoC are given as input including the two following
elements:

1. *Target List of Initiator Module* (TL_IM). For each initiator, it includes a list of
   target modules allowed to access.
2. *Secure Target List* (STL), including the target modules containing critical data.

The formal definition of design specification in the *design security rules* file is as follows:

Definition 5.2

$$ST L = \{t_i \mid t_i \ is \ a \ secure \ memory \ ; \ 1 \le i \le n\}$$

$$TL\_IM = \{IM_i \mid IM_i \rightarrow \{t_j \ ; \ 1 \le j \le m\} \ ; \ 1 \le i \le n_I\}$$

where n and m illustrate the number of secure target module and the number of targets that initiator module IM_i is allowed to access, respectively. The parameter n_I indicates the number of initiator modules of the design.

Algorithm 5.1 illustrates the algorithm of generating *Direct Security Property* (DSP) from the *design security rules*. First, the *TL_IM* list is analyzed to extract all target modules of the design and store them in the *Design Target List* (DTL) (line 1). For each initiator module *IM* of the design a *Forbidden Target List* (FTL_{IM}) is generated containing the target modules that the initiator is not allowed to access (lines 4 and 5). This is performed by eliminating the *Target List* (TL) of initiator *IM* (defined in *TL_IM*) from DTL. Finally, for each target in *FTL*, a property is generated including the initiator name, target name, and the access mode (lines 6 to 9). Each direct security property p_i in DSP identifies that there must be no flow (e.g., read or write access) from initiator *IM* to target T (lines 7 and 8).

The generation of the *Indirect Security Property* (ISP) is explained in Algorithm 5.2. Similar to Algorithm 5.1, it starts with extracting the design's target modules from the *TL_IM* set (line 1). In order to detect the indirect scenario, the generated ISP must include three sequences as follows:

- the first sequence (seq_1) indicates a read access from a secure initiator (i.e., the initiator module allowed to access a secure memory) to its secure target (lines 4 to 7),
- the second sequence (seq_2) identifies a write access from the secure initiator of seq_1 to a non-secure memory (lines 8 and 9), and

---

Algorithm 5.1: Direct security property generation

Data: targets' lists of initiator modules *TL_IM*
Result: Security properties to validate direct acess paths *DSP*
1 $DTL \leftarrow$ extracting all target modules of design form *TL_IM*;
2 $i \leftarrow 0$;
3 $DSP \leftarrow \emptyset$;
4 **foreach** *target list TL of initiator module I M in TL_IM* **do**
5       $FTL_{IM} \leftarrow DTL - TL$;
6   **foreach** *target T in FTL_{IM}* **do**
7           $p_i \leftarrow (IM, T, \{R/W\})$;
8           $add(DSP, p_i)$;
9           $i \leftarrow i + 1$;

Algorithm 5.2: Indirect security property generation

Data: Secure target list *STL* and targets' lists of initiator modules *TL_IM*
Result: Security properties to validate indirect acess paths *ISP*
1   *DTL* ← extracting all target modules of design form *TL_IM*;
2   *i* ← *0*;
3   *ISP* ← ∅;
4   **foreach** *target module T in STL* **do**
5 | **foreach** *initiator module IM in TL_IM* **do**
6 | | **if** *T in target list TL of IM* **then**
7 | | | seq_1 ← $(IM, T, \{R\})$;
8 | | | **foreach** *target module T′ in DTL − STL* **do**
9 | | | | **if** *T′ in target list TL of IM* **then**
10 | | | | | seq_2 ← $(IM, T', \{W\})$;
11 | | | | | **foreach** *initiator module IM′ in TL_IM − IM* **do**
12 | | | | | | **if** *T′ in target list TL of IM′* **then**
13 | | | | | | | seq_3 ← $(IM', T', \{R\})$;
14 | | | | | | | p_i ← (seq_1, seq_2, seq_3);
15 | | | | | | | *add(ISP, p_i)*;
16 | | | | | | | i ← $i + 1$;

- the third sequence (seq_3) is a read access from a non-secure initiator (i.e., the initiator module that is not allowed to access a secure memory) to the target of seq_2 (lines 10 and 11).

Consider the running example *RISC-32-SoC* (Fig. 3.1). The *design security rules* file is defined as follows:

$$STL = \{M2\_secure\} \tag{5.1}$$

$$TL\_IM = \{Initiator\_0 \rightarrow \{M0\_regular, M1\_shared\},$$

$$RISC\_CPU \rightarrow \{M1\_shared, M2\_secure\}\}$$

Based on its *design security rules*, the *DSP* and *ISP* are generated as follows:

$$DSP = \{p_1, p_2 \mid p_1 \rightarrow (Initiator\_0, M2\_secure, \{R/W\}), \tag{5.2}$$

$$p_2 \rightarrow (RISC\_CPU, M0\_regular, \{R/W\})\}$$

$$ISP = \{p_1 \mid p_1 \rightarrow ((RISC\_CPU, M2\_secure, \{R\}), \tag{5.3}$$

$$(RISC\_CPU, M1\_shared, \{W\}),$$

$$(Initiator\_0, M1\_shared, \{R\}))\}$$

Please note that when a part of memory (range of memory addresses) is defined as the secure part, the *design security rules* can be modified to support it.

Algorithm 5.3: Security violation detection

Data: set of access paths *AP* and security properties *DSP* and *ISP*
Result: violated access paths V_{AP}
1 /* validating AP against direct security properties   */
2 **foreach** *direct property dp in DSP* **do**
3   **foreach** *path p in AP* **do**
4   **if** *(dp in p)* **then**
5              $add(V_{AP}, p)$;

6 /* validating AP against indirect security properties   */
7 **foreach** *indirect property ip in ISP* **do**
8   **foreach** *path p in AP* **do**
9   **if** *(ip.seq$_1$ in p)* **then**
10   **foreach** *path p' in AP* **do**
11   **if** *(ip.seq$_2$ in p')* & *(p.ST > p'.ST)* **then**
12                    $add(V_{AP}.suspected, (p, p'))$;
13   **foreach** *path p'' in AP* **do**
14   **if** *(ip.seq$_3$ in p'')* & *(p'.ST > p''.ST)* &
                                  (ip.seq$_2$.adrs == ip.seq$_3$.adrs) **then**
15                          $add(V_{AP}, (p, p', p''))$;

5.4.5 Security Validation

In the case of *direct access*, for each property in *DSP* the *AP* set is traversed (Algorithm 5.3—lines 1 to 4) in order to find a security policy violation. The detection of a property violation in this case signifies that there is a direct path between a non-secure initiator module and a secure memory. All violated paths are stored in *Violated Access Path* (V_{AP}) set to be reported to designers. To detect a property violation related to *indirect access* (lines 5 to 13), the *AP* set of design is analyzed to find a combination of direct access paths in which the whole or a part of secure data is leaked. We split the security violation in this case into two cases which are *suspected* and *violated*.

The former refers to the two sequential accesses of an initiator where it reads a sensitive data from a secure memory and then writes the data to a non-secure memory. We define these two accesses as a suspect case as it may lead to an information leakage. The *suspected* case is detected when the first two sequences (seq_1 and seq_2) of an indirect security property are violated (lines 7 to 10). The paths related to this violation are stored in $V_{AP}.suspected$ set. When a non-secure initiator reads the sensitive data from the non-secure memory written by the secure initiator (lines 11 to 13), the *suspected* case becomes a real indirect security violation that is defined as *violated* case. Therefore, the analysis reports the paths of violated properties in V_{AP} set, e.g., verifying the *AP* set of the running example *RISC-32-SoC* (Fig. 5.4) against its security properties shows that an ISP is violated. The access paths reported by the analysis illustrate that the secure data is leaked by combination of direct paths $\{P2, P4, P5\}$.

To know how much of the secret information is leaked (considered as leakage depth analysis), the data length attribute (L) of the transaction specifies the size of data that is read from or written to the memory block by the initiator module. In case that the violated property is a DSP, parameters *adrs* and L (in the related path) signify the depth of leakage. Regarding *indirect access*, the analysis must be performed on more than one path as the violated ISP includes three sequences. The access paths of the last two sequences seq_2 and seq_3 contain the information of writing the secure data to and reading it from a non-secure memory, respectively. The *adrs* parameter of both sequences identify the accessed block of memory by the initiators. If data length in seq_3 (L_3) is equal or smaller than data length in seq_2 (L_2), the depth of leakage is equal to L_3 otherwise is L_2.

5.5 Experimental Evaluation

The experimental evaluation of the proposed approach has been performed on various VP models implemented in SystemC TLM-2.0. The experiments cover both the generality and scalability of the approach. The former refers to the security validation of designs [6] implementing various aspects of the TLM-2.0 standard (core-interfaces, the base protocol, and coding styles). The latter refers to the security validation of a real-world VP-based SoC [101].

To evaluate the quality of the proposed approach, we consider two possible security scenarios covering both confidentiality and integrity threat models:

- S1: modifying an SoC without updating its security policies (similar to the motivation example scenario in Sect. 5.3) and
- S2: incorrect initialization or update of the SoC firmware (i.e., memory configuration file).

We apply the aforementioned security scenarios to VP models and verify their security policies against their security rules.

The experimental results are described in two parts. First, a real-world case study—the LEON3-based VP SoCRocket (implemented in SystemC TLM-2.0) [101] is illustrated in detail in Sect. 5.5.1. Second, we give a brief discussion on the quality of obtained experimental results in Sect. 5.5.2.

All the experiments are carried out on a PC equipped with 8 GB RAM and an Intel Core i7 CPU running at 2.4 GHz.

Table 5.1 Experimental results for all case studies related to the quality of the proposed security validation approach

VP model[a]	LoC	TM	SSc	#IP			#Trans	#ISP			#DSP		
				Total	SI	SM		Total	Pass	Fail	Total	Pass	Fail
Routing-model (1)	656	LT	S1	9	2	2	150	24	20	4	5	4	1
RISC32-SoC (1,2)	3150	LT	S1	6	1	1	110	1	0	1	2	2	0
AES128-SoC (1)	4742	AT	S1,S2	10	2	3	350	33	31	2	13	8	5
RISC32-SoC (1,2)	4850	AT	S1	19	3	3	733	324	304	20	36	34	2
Locking-two (1)	5830	LT/AT	S2	13	2	2	450	48	39	9	14	10	4
Locking-auto (1)	6959	LT/AT	S2	15	2	3	370	96	85	11	20	18	2
SoCRocket (3,4)	50,000	LT/AT	S2	21	1	3	1100	180	168	12	30	28	2

[a]The VP models are provided by (1) [6], (2) [1], (3) [101], and (4) [97] and modified using a different combination of initiator and target modules to support various security scenarios
LOC lines of code, *TM* timing model, *SSc* security scenario, *#IP* number of intellectual property, *SI*, secure initiator, *SM* secure memory, *#Trans* number of transactions, *ISP* indirect security property, *DSP* direct security property

5.5.1 Case Studies

The experimental results for different types of SoC benchmarks are shown in Tables 5.1 and 5.2. The first two columns in Table 5.1 list the names and lines of code for each VP model, respectively. Note that the original model of each SoC is modified by integrating different initiator and target modules with its interconnect to support various security scenarios. Column *TM* presents the timing model of each design. Column *SSc* shows the security scenario applied to each VP model. Column *#IP* lists the number of IPs for each VP model followed by the total number of IPs, secure initiator, and secure memory in *Total*, *SI*, and *SM*, respectively. The column *#Trans* shows the number of extracted transactions for each design. Columns *#ISP* and *#DSP* present the result of verifying each VP model against the indirect and direct security properties, respectively. For both columns, *Total*, *Pass*, and *Fail* illustrate the number of generated, satisfied, and violated properties, respectively. The execution time of the proposed approach is reported in Table 5.2, including the data extraction *ET*, validation process *VT*, and the total execution time *Total* of the approach. The time reported in *VT* covers phases 2 to 4 of the proposed approach. Column *CET* shows the total time of each design's compilation and execution without any instrumentation (Table 5.2).

For a real-world experiment, we modified the LEON3-based VP SoCRocket [101] by integrating three TLM-2.0 IPs with its AMBA-2.0 AHB (Advanced High-performance Bus) as the following:

- A synthesizable SytemC IP *AES_core* [97] used as a hardware accelerator to implement AES-128 encryption algorithm.
- Two secure memories *Mem_Secure1* and *Mem_Secure2* initialized by *cryptography keys* and *plain texts*, respectively.

Table 5.2 Experimental results for all case studies related to the execution time of the proposed security validation approach

VP Model[a]	LoC	TM	SSc	#IP			#Trans	ET (s)		VT (s)	Total (s)		CET (s)
				Total	SI	SM		DbA	CbA		DbA	CbA	
Routing-model (1)	656	LT	S1	9	2	2	150	41.06	2.57	15.72	56.78	18.29	1.52
RISC32-SoC (1,2)	3150	LT	S1	6	1	1	110	35.39	4.39	4.92	39.06	9.31	2.78
AES128-SoC (1)	4742	AT	S1,S2	10	2	3	350	208.15	20.06	34.90	243.05	54.96	16.23
RISC32-SoC (1,2)	4850	AT	S1	19	3	3	733	556.93	21.53	52.23	609.16	73.76	23.84
Locking-two (1)	5830	LT/AT	S2	13	2	2	450	402.72	23.67	29.11	431.83	52.78	22.21
Locking-auto (1)	6959	LT/AT	S2	15	2	3	370	513.40	24.81	39.82	553.22	64.63	22.70
SoCRocket (3,4)	50,000	LT/AT	S2	21	1	3	1100	1556.49	59.18	104.07	1660.56	163.25	43.19

[a]The VP models are provided by (1) [6], (2) [1], and (3) [101] (4) [97] and modified using a different combination of initiator and target modules to support various security scenarios

LOC lines of code, *TM* timing model, *SSc* security scenario, *#IP* number of intellectual property, *SI* secure initiator, *SM* secure memory, *#Trans* number of transactions, *ET* extraction time of the proposed approach (phase 1), *VT* validation time of the proposed approach (phases 2 to 4), *CET* compilation and execution time of each VP model using GCC

The VP itself is implemented in SystemC TLM-2.0 consisting of more than 50,000 lines of code. It includes several IPs working together in master (e.g., initiator modules *LEON3* processor, *ahbin1*, and *ahbin2*) or slave (e.g., target modules *AHBMem1* and *AHBMem2*) mode which are connected to the on-chip bus AMBA-2.0. The communication uses a 32-bit address mode where the 12 most significant bits are used to specify the memory address.

The expected security policies of the VP are as follows:

- *Mem_Secure1* and *Mem_Secure2* are secure memories and only accessible by *AES_core* and
- *AHBMem1* and *AHBMem2* are regular memories and only accessible by *LEON3*, *ahbin1*, and *ahbin2*.

Initially, memory *AHBMem2* is not available and memory configuration is defined based on the aforementioned security policies. At the beginning of execution, initiator modules read the memory configuration file to extract the range of memory addresses that they are allowed to access. The *AES_core* module executes the standard AES-128 encryption algorithm using the *initialized keys* and *plain texts* stored in secure memories *Mem_Secure1* and *Mem_Secure2*, respectively.

In order to increase the overall performance of the system, consider the scenario that the design team decides to integrate *AHBMem2* with the *AT-bus* (AHB). In order to make the new memory accessible by other initiators, the memory configuration file needs to be updated. The expected update from the design team for memory configuration is as follows: in *AHBMem2*, memory blocks

- (0xA0000000 to 0xA0000BB4) are shared among *LEON3*, *ahbin1*, and *ahbin2*,
- (0xA0000BB5 to 0xA0000DE6) are only accessible by *AES_core*, and
- (0xA0000DE7 to 0xA0000FFF) are shared between *ahbin1* and *ahbin2*.

The security scenario is that the memory configuration file is incorrectly updated (either by malicious insider on the design team or unintentionally) where in *AHBMem2*, memory blocks

- (0xA0000000 to 0xA0000BB4) are shared among *LEON3*, *ahbin1* and *ahbin2*,
- (**0xA0000BA5** to 0xA0000DE6) only accessible by *AES_core*, and
- (0xA0000DE7 to 0xA0000FFF) are shared between *ahbin1* and *ahbin2*.

The wrong memory configuration in this case is difficult to be detected by conventional functional tests because of two reasons: First, this fault does not affect the functionality of the system as the location of storing variables is only shifted in the same memory. Second, from the functional point of view, *AHBMem2* is reachable by all expected initiators and the transactions data is still stored in the expected memory (*AHBMem2*). From the security point of view, the secure memories *Mem_Secure1*, *Mem_Secure2*, and *AHBMem2* (0xA0000BB5 to 0xA0000DE6) must be protected against unauthorized access. Hence, the *Design security rules* of the *SoCRocket* are defined as the following:

$$ST L = \{Mem\_Secure1, Mem\_Secure2, AHBMem2 \ (0xA0000BB5-$$

$$0xA0000DE6)\}$$

$$TL\_IM = \{AES\_core \rightarrow \{Mem\_Secure1, Mem\_Secure2,$$

$$AHBMem2 \ (0xA0000BB5 - 0xA0000DE6)\},$$

$$LEON \rightarrow \{AHBMem1, AHBMem2 \ (0xA0000000 - 0xA0000BB4)\},$$

$$ahbin1 \rightarrow \{AHBMem1, AHBMem2 \ (0xA0000000 - 0xA0000BB4;$$

$$0xA0000DE7 - 0xA0000FFF)\},$$

$$ahbin2 \rightarrow \{AHBMem1, AHBMem2 \ (0xA0000000 - 0xA0000BB4;$$

$$0xA0000DE7 - 0xA0000FFF)\}\} \tag{5.4}$$

The proposed approach generates 210 security properties (30 DSPs and 180 ISPs) w.r.t the *design security rules*. It detects 16 security properties violation including two DSPs and 14 ISPs. The main reason for this security problem is the weak security policy of AMBA-2.0 AHB. The only policy implemented in AMBA-2.0 AHB is that for receiving transactions generated by master IPs, it checks whether or not the transactions' address is in the range of memory addresses. We fixed this security gap in AMBA-2.0 AHB by adding access control policies restricting the access of unauthorized master IPs to the secure memories. This can be done by checking the address of the received transactions whether or not they satisfy the expected range of addresses w.r.t the design security rules. The properties were satisfied on the next analysis run. A more general solution is to add a memory management unit to the AMBA-2.0 AHB. For instance, the V_{AP} set related to the violated DSP is as follows:

$$V_{AP} = \{P14, P21 \ |$$

$$P14 = (0x69C450\_4 \rightarrow ahbin1 :: init1 :: gen\_frame \rightarrow READ$$

$$\rightarrow AHBMem2 :: trg2 :: exec\_func \rightarrow 0xA0005BA7 \rightarrow 4 \rightarrow 6071ns),$$

$$P21 = (0x6A9B20\_2 \rightarrow ahbin2 :: init2 :: gen\_frame \rightarrow READ$$

$$\rightarrow AHBMem2 :: trg2 :: exec\_func \rightarrow 0xA0005BB1 \rightarrow 4 \rightarrow 8043ns) \tag{5.5}$$

Path $P14$ in (5.5) shows that instance *init1* of initiator module *ahbin1* generates a transaction using *gen_frame* function to read from memory address 0xA0005BA7 referring to instance *trg2* of target module *AHBMem2*. The target module *AHB-Mem2* sends the response using *exec_func* at simulation time 6071 ns. In a same way, path $P21$ demonstrates that instance *init2* of initiator module *ahbin2* generates a transaction using *gen_frame* function to read from memory address 0xA0005BB1 referring to instance *trg2* of target module *AHBMem2*. The target module *AHBMem2* sends the response using *exec_func* at simulation time 8043 ns.

The leakage depth analysis on V_{AP} in (5.5) shows that for both paths $P14$ and $P21$, the parameter $adrs$ is equal to 0xA0005BA7 and 0xA0005BB1, respectively. The parameter L is equal to 4 Bytes for all paths.

A part of the V_{AP} set regarding the ISP is as follows:

$$V_{AP} = \{(P411, \ P532, \ P565) \ |$$

$$P411 = (0x1DBCD00\_21 \rightarrow AES\_core :: aes\_master :: gen\_frame$$

$$\rightarrow READ \rightarrow Mem\_Secure1 :: mem\_sec1 :: exec\_func$$

$$\rightarrow 0xB0000010 \rightarrow 4 \rightarrow 999550ns),$$

$$P532 = (0x1DBCD00\_54 \rightarrow AES\_core :: aes\_master :: gen\_frame$$

$$\rightarrow WRITE \rightarrow AHBMem2 :: trg2 :: exec\_func$$

$$\rightarrow 0xA0000BA9 \rightarrow 4 \rightarrow 1007190ns)$$

$$P565 = (0x6A9B20\_19 \rightarrow ahbin2 :: init2 :: gen\_frame$$

$$\rightarrow READ \rightarrow AHBMem2 :: trg2 :: exec\_func\_func$$

$$\rightarrow 0xA0000BA9 \rightarrow 4 \rightarrow 1008680ns)\} \qquad (5.6)$$

As illustrated in (5.6), the combination of paths $P411$, $P532$, and $P565$ creates an indirect data flow between $ahbin2$ and $Mem\_Secure1$, which is against the expected security rules. In path $P411$ instance aes_master of initiator module AES_core creates a transaction to read from memory address 0xB0000010 referring to the secure target module $Mem\_Secure1$ using gen_frame function. In path $P532$ the AES_core generates a transaction to write in memory address 0xA0000BA9 referring to target module $AHBMem2$ when the simulation time advanced from 999550 to 1007190 ns. Finally, in path $P565$ instance $init2$ of initiator module $ahbin2$ reads from memory address 0xA0000BA9 at simulation time 1008680 ns.

For this experiment, the number of extracted transactions are 1100 and the whole analysis takes about 28 min to report the results.

5.5.2 Integration and Discussion

The proposed approach is able to verify the security of abstract communication for a given VP-based SoC model. The result of our analysis can effectively guide the verification engineer to improve the security policies of the design as it reports the exact vulnerable paths. Since the first phase of the proposed approach (information extraction) can be performed either the debugger-based or compiler-based approaches, designers have this option to choose each of them based on their requirements. The debugger-based approach only requires the binary VP models for IFT, thus the original source code and compilation workflow stay untouched (i.e., a non-intrusive solution). Thus, it can be used in case that the SoC includes 3PIPs where their source codes are not available. The only precondition for the application of the suggested approach is that the executable VP model contains debug information which is normally provided by vendors. The compiler-based approach provides designers with a fast solution for information extraction; however, it requires the availability of the SoC source code.

5.5.2.1 Limitations

As the proposed approach is based on dynamic analysis, it inherits the same limitations. The validation analysis depends on the ability of the input stimulus (i.e., testbench or running software) to activate the vulnerabilities or attacks surface in the first phase. The probability of the vulnerabilities activation in the design's simulation behavior can be increased based on the fact that each initiator module of the design accesses different target IPs (e.g., memory address ranges) for which they are defined at least with one input stimulus (transaction). This is a good starting point to generate input stimulus or to improve the existing testbench for a given design as various combinations of paths appear in the simulation log. High quality of input stimulus (that is free of redundant paths) can also reduce the data extraction time (phase 1) as the number of transactions that need to be traced is reduced. Generally, the problem of generating input stimulus or improvement on the existing one can be solved by either using automated test generation methods or manually exercising a specific path.

5.6 Conclusion

In this chapter, the first dynamic VP-based IFT approach for security validation was introduced. The IFT is performed by utilizing the suggested SystemC VP analysis approaches (i.e., compiler-based and debugger-based) to extract the run-time behavior (TLM transactions) of a given SoC. The extracted transactions and the design security rules are automatically translated into a set of access paths and security properties, respectively. The proposed approach validates the generated access paths against the security properties and reports back the vulnerable paths (that violate specified confidential information flow properties) to the designer for further inspection. The proposed approach is able to detect the exact point and amount of information that is leaked. Experimental results confirmed the applicability of our approach on various VP model, including the real-world VP SoCRocket.

Chapter 6
Application III: Design Space Exploration

The increasing functionality of electronic systems due to the constant evolution of the market requirements makes the non-functional aspects of such systems (e.g., energy consumption, area overhead, or performance) a major concern in the design process. Approximate computing is a promising way to optimize these criteria by trading accuracy within acceptable limits. Since the cost of applying significant structural changes to a given design is very expensive and increases with the stage of development, the optimization solution needs to be incorporated into the design as early as possible, i.e., at the ESL. In order to apply approximation techniques to optimize a given SystemC design, designers need to know which parts of the design can be approximated. However, identifying these parts is a crucial and non-trivial starting point of approximate computing as the incorrect detection of even one critical part as resilient may result in an unacceptable output. This usually requires a significant programming effort by designers, especially when exploring the design space manually.

In this chapter, we present another application of the design understanding in the design process, which is identifying the resilient portions of a given SystemC design. To do so, we use the compiler-based analysis approach (introduced in Chap. 3) along with regression analysis techniques (a fast machine learning method providing an accurate function estimation). Once the resilient portions are identified, an approximation degree analysis is performed to determine the maximum error rate that each resilient portion can tolerate. Moreover, the maximum number of resilient portions that can be approximated at the same time is reported to designers at different granularity levels.

© Springer Nature Switzerland AG 2020
M. Goli, R. Drechsler, *Automated Analysis of Virtual Prototypes at the Electronic System Level*, https://doi.org/10.1007/978-3-030-44282-8_6

6.1 Introduction

Due to the ever-increasing functionality of digital circuits and the constant evolution in market requirements for silicon solutions, the final electronic systems have become very complex. This rising complexity comes at a price such as greater power consumption, area overhead, or performance reduction. Hence, the aforementioned non-functional design aspects have become a severe concern for designers of such complex systems.

Approximate computing paradigm is a promising solution [23, 77] to improve the performance, area, or to reduce the required energy consumption of electronic systems at the cost of output accuracy. Based on the idea that designs usually include some parts that contribute to the quality of output less than others, modifying these parts of the design can enhance design metrics. The output quality is measured based on its acceptable error boundary that is specified as a Quality of Service (QoS) range. The inaccuracy toleration or QoS bound of the final output of a given design can arise due to different factors such as the inability of humans to perceive noise within limits. The parts of a given design are considered as resilient or *approximable* if they do not affect the final output beyond a tolerable limit. Thus, the modification of the approximable parts has a low impact on the output's QoS.

Since the cost of applying significant structural modifications to a given design increases with the stage of development, designers must craft their solutions to be incorporated into the design as early as possible, i.e., at the ESL using SystemC language. Moreover, a significant impact on overall design metrics can be only achieved when the modifications are applied at the high levels of abstraction during the design process.

However, before any approximate computing technique can be applied to a given SystemC design, the approximable parts of the design need to be identified. This is the critical starting point of approximate computing as the incorrect detection of critical parts as approximable can be catastrophic in terms of the output quality of the design.

According to [42, 77], a solution that locates the resilient portion of a design (which identifies the most promising approximation candidates) should be as automated as possible. The existing solutions are mostly based on developing new programming languages to provide designers with frameworks to manually specify approximable data [72, 95] or source code annotations to determine whether or not a part of the code is resilient [14, 15, 17, 72, 91]. Moreover, the methods focus on either algorithmic level [14, 84, 91–93] or lower levels of abstraction that is the *Register Transfer Level* (RTL) [83] and below [16, 96].

Hence, in this chapter, we aim to automatically identify the approximable portions of a given SystemC design, that can, in turn, be realized and mapped onto the approximate implementations by designers in the later design steps. The proposed approach consists of four main phases. In the first phase, a combination of static and dynamic analysis techniques is used to extract the simulation behavior of different design portions. In the second phase, regression analysis techniques are applied to the extracted simulation behavior of each portion to find a real approximated model with negligible error. In the third phase, the original portion is replaced with its estimated model (that has very close behavior to its true behavior) to evaluate its impact on the final output of the design. If the QoS of the design is satisfied, the portion is marked as approximable. In the last phase, to help designers to make better decisions on design approximation and subsequently a better optimization on different aspects of the design, an approximation degree analysis is performed. The analysis is done at different granularity levels starting from an approximable portion and continues towards the entire design (a bottom-up analysis). The results of this analysis provide designers with the maximum

- error rates that an approximable portion can accept,
- number of module's portions that can be approximated at the same time, and
- number of modules that can be approximated at the same time w.r.t the QoS of the design.

The proposed approach is applied to several standard SystemC case studies, including a real-world JPEG-encoder VP, to show its advantages such as precision and scalability.

The rest of this chapter is organized as follows. Sect. 6.2 outlines the existing methods in this area and compares them to the proposed approach. In Sect. 6.3, we introduce the proposed approach. The experimental results are discussed in Sect. 6.4. Finally, the chapter is concluded in Sect. 6.5.

6.2 Related Works

In this section, we discuss relevant research in approximate computing related to finding the approximable portions of a program at different levels of abstraction.

Approximate computing has been applied to the design of hardware domain modeled at RTL and below such as arithmetic units [47], data paths [67], or voltage scaling [78]. The leverage of machine learning algorithms to improve the energy consumption has been introduced in [26, 113] where the computational parts of the design are replaced with hardware neural networks. However, their applications are limited to some particular designs. Moreover, the impact of using such techniques is significant if they are applied at the higher levels of abstraction.

Several approaches [14, 35, 118] have been introduced related to classifying program code (or its underlying structures) as either critical or approximable. They mostly analyze the sensitivity (the degree of an output's volatility in relation to distortions in each variable in a program) based on statistical techniques [84, 93] in order to specify the approximable variables at the algorithmic level.

In [14], a combination of dynamic analysis and data mining techniques is used to specify the critical inputs and the corresponding code regions of a given program. First, the program is run with a set of illustrative inputs and the baseline executions are recorded. The program is then executed with distorted inputs and the behavior differences are compared to the baseline execution. A post-execution analysis is performed using a data mining technique to classify inputs and the corresponding code regions as either approximable or non-approximable. However, the method does not provide any information regarding the classification of program internal data.

In [93], a statistics-based method is introduced to locate approximable structures in a program automatically. First, variables of the program and the range of values are extracted. Then, during the "error injecting phase," the value of each variable is perturbed to measure the effect of the modification on the output. If the new output lies within the reference QoS threshold, the variable is marked as approximable. In [84], the same technique is used to analyze the sensitivity of each variable in a program. The difference is that it does not perform a range analysis of variables as it takes advantage of memory bit flips for its error injection phase. Moreover, the number of program executions in the presence of perturbations to perform sensitivity analysis is selected using probabilistic computations while in [93] it is set manually.

The method in [91] presents a sensitivity analysis based on a probabilistic model by partitioning a computation into tasks and identifying the resilient ones in such a way that abandoning those do not affect the output w.r.t its acceptable QoS. The probabilistic model is obtained by sampling the program execution randomly at different tasks failure versus the QoS failure rates. Then, this model is used to identify resilient tasks. However, the method requires designers to partition the computation into tasks manually.

Similarly, [92] performs a sensitivity analysis in parallel computing. For a given computation (partitioned into several tasks), it performs an early phase termination at barrier synchronization points. Then, the effect of terminating tasks, early to the computation result, is analyzed using probabilistic models. Tasks in computations which are tolerant to early phase terminations are marked as resilient.

Recently, [76] presents a static significant analysis approach to identify parts of a program that can be approximated. It is based on the path extraction analysis to express symbolically the output only with variables which are specified by designers. Then, it ranks the variables based on their effect on the output. Although the method is able to rank the importance of variables (which needs to be defined by designers manually) associated to the program output, it does not provide designers with any information about the approximation degree of variables. Thus, it can only be used to prune the variables with high importance from the entire candidates for approximation.

However, none of the aforementioned methods can be applied to hardware systems easily as they disregard any timing and architectural information. Moreover, the performance of all methods that perform resilience evaluation at the granularity level of a local data is related to the number of variables and the number of samples (which corresponds to the required program executions). Therefore, the execution time increases with the number of variables, leading to significant issues with larger programs that rely on local data.

Some works have been proposed [15, 72, 95] based on source code annotations and type qualifiers to indicate whether a variable or data is approximable. For example, [95] provides designers with an extension to the Java programming language where designers can annotate data types to be either precise or approximate. However, using the aforementioned methods require designers to rewrite or annotating source codes manually. Even for a simple model, it requires lots of programming effort, learning the new language constructs and overall, very time-consuming process.

The motivation of this work is to present a novel analysis approach that eliminates the drawbacks of the existing solutions which are in short: (1) lack of supporting designs at the ESL, (2) significant execution time overhead, (3) low degree of automation, and (4) lack of approximation degree analysis.

6.3 Methodology

In this section, we first give a top-level overview of the proposed approach. Then, we explain its four phases in detail.

6.3.1 Overall Workflow

As demonstrated in Fig. 6.1, the process of finding resilient portions of a given ESL design is divided into four main phases:

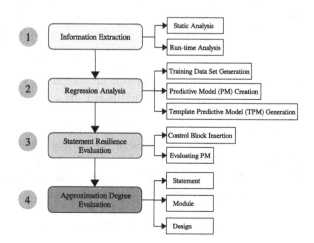

Fig. 6.1 The proposed approach overview

1. Information extraction, including two steps:

   • analyzing the AST of the design to extract the computational statements and
     data dependency graph, and
   • generating an instrumented model of the design and executing it to retrieve
     run-time information.

2. Performing regression analysis techniques on the extracted run-time information
   based on the following steps:

   • translating the extracted run-time information into the training data set,
   • creating a *Predictive Model* (PM) to estimate the behavior of the extracted
     data, and
   • generating a *Template Predictive Model* (TPM) from the PM for approxima-
     tion degree evaluation.

3. Evaluating the resilience of each design's portion by replacing the original
   portion with its corresponding generated PM.
4. Finding the maximum

   • error rate that each approximable statement can accept,
   • number of statements that can be approximated in a module, and
   • number of modules that can be approximated in the design w.r.t the design's
     QoS.

6.3.2 Information Extraction

The behavior of a given SystemC design is specified by a set of modules that
communicates through signals. Each module includes one or more processes or
functions to perform different tasks representing the module's and overall design's
behavior. To perform a task, each process or function includes a set of *computational*
and *control* statements. The former refers to the calculation of a finite number
of input values (considered as operands) to an output value. The latter refers to
the order in which the computational statements are executed, such as conditional
(e.g., if-statement or switch-statement) or loop (for-loop or while-loop) statements.
As stated in [77], in the approximate computing paradigm, the approximation
can only be applied to the non-critical parts of a given design as approximating
critical parts, e.g., control statements can lead to catastrophic consequences, such as
segmentation fault, program crash, or erroneous output. Hence, to find the resilient
portions of a given SystemC design, the starting point of our analysis is to evaluate
the portions of the design, which essentially are not prohibited and have a higher
chance of being approximated, i.e., the computational statements. Thus, the first
step is to find the computational statements of the design. Once these statements
are detected, the run-time information related to their behavior is extracted. To do
this, we take advantage of the compiler-based approached (introduced in Chap. 3,

Sect. 3.4.2) to statically analyze the AST of the design and generate its instrumented model for run-time information extraction.

6.3.2.1 Static Analysis

The first step of the static analysis is to find the computational statements. To do this, we need to know (1) how a computation statement is defined w.r.t the SystemC constructs and (2) which computation statements need to be extracted in a given SystemC design. The computational statements are usually defined as assignments in the design. However, it is possible to define a computation statement as a return value of a function call. Moreover, due to the SystemC structure, it is also possible to assign a computation of some variables to the output ports of a module. This is done by calling the *write()* member function of the *sc_out* class of the SystemC library. We only extract the information related to the computation statements that have at least two operands. Thus, the simple statements such as $a = 0$ or $a = b$ are not considered as our target statements. The reason is that the left-hand side variable of these simple statements appears later in a complex statement (either a computational or control statement). In the case of a computational statement, its resilience is evaluated when the complex statement is evaluated. Moreover, this reduces the number of statements that need to be analyzed for resilience evaluation in the later steps and overall decrease the required analysis time. We do not consider the computational statements that are defined in the condition of a control statement (e.g., *if* or *else if*) as they directly affect the control flow of the design.

In the case of an assignment statement, the left-hand side operand can be either a global or local variable (signal). In addition to this, the right-hand side operand can be a function call or an input port. Regarding the *write()* member function of the output ports and the return statement of a function, both can receive an expression like the right-hand side of an assignment statement. If a statement includes a function call, we consider the function as an operand of the statement and define the function name as the operand name. Moreover, concerning the return statement of a function, in addition to the operands of the statement, a new operand with the name of the function is defined to hold the value of the return statement. Therefore, all the target computational statements of a given SystemC design can be specified using the following definition.

Definition 6.1 A statement S_i is a tuple (I, R) where I is the set of input operands and R is the result of S_i.

$$S_i = \{(I, R) \mid R \sim I, \ I = \{op_1, op_2, \ldots op_n\}\}$$

where i shows the line of code where the statement S is defined and n is the number of its input operands. Since the result value of a target statement (e.g., the left-hand side variable of an assignment statement) may be defined differently in the body of a control block (e.g., *if-else* or *switch-case* statement), the line of code where the statement is defined is used as the distinguishing parameter.

For example, consider the motivating example (Fig. 3.2) which illustrates a SystemC design including two modules *M1* and *M2*. The design performs a set of algebraic operations in two stages to generate the final results. Now consider lines 33, 41, and 49 of the design, which are a member function call of the output signal *out2*, a member function call of the output signal *out1* including a function call, and a return statement, respectively. The specifications of these statements based on definition 6.1 are as the following:

$$S_{32} = \{(I, out2) \mid out1 \sim I, \ I = \{temp2, temp1\}\} \tag{6.1}$$

$$S_{41} = \{(I, out1) \mid tp1 \sim I, \ I = \{mult, tp1\}\}$$

$$S_{49} = \{(I, mult) \mid mult \sim I, \ I = \{in1, in3\}\}$$

The target statements are detected by visiting relevant nodes in the AST of the design. We consider these locations as the *Target Statement* (TStm) locations.

After detecting *TStm* locations, the next step is to extract the structural information of the *TStms* in order to identify their relationship. The first step of extracting the statements relationship is to find all statements with the same result variable that are defined in the body of a conditional block (e.g., *if-else* or *switch-case*). These statements are categorized in the same group. This reduces the number of statement combinations for the approximation degree analysis (the fourth phase of the proposed approach). The name of each group is defined based on a hierarchical structure including

- the name of module,
- the name of function, and
- the line of code where the conditional block is defined.

For example, the result of applying the statement dependency analysis to the module *M1* of the motivating example (Fig. 3.2) shows that S_{24} (line 24) and S_{27} (line 27) are categorized in the same group with the name of condition block and the line of code as the following:

$$G : M1\_func1\_if\_23 = \{S_{24}, S_{27}\} \tag{6.2}$$

The second step of the statements relationship extraction is to generate a data dependency graph w.r.t the following definition.

Definition 6.2 The data dependency graph is a structure (N, E, Z), where N is a set of nodes, E is a set of edges, and $Z \subseteq N$ is set of output variables. The edge from node X to node Y shows that Y is dependent to X.

Each node of the data dependency graph is a variable of the design which is tokenized by the name of module and function (for local variable) to which the variable belongs. This graph is required to identify those statements which contribute to

Fig. 6.2 Data dependency graph of the motivating example. The gray nodes represent the result variable of target statements

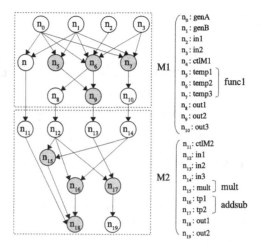

generating the output of the design which is supposed to be approximated. By this, the resilience evaluation is only performed on the target statements that are related to this output.

For example, Fig. 6.2 illustrates the data dependency graph of the motivating example (Fig. 3.2). The gray nodes in this figure represent the variables which are the results of the target statements. Nodes n_{18} and n_{19} show the final output of the design. If the output signal *out2* of the *M2* module (node n_{19} in Fig. 6.2) is the only final output of the design that is supposed to be approximated by designers, the resilience evaluation needs only to be applied to the target statements associated to *out2*. In this respect, the resilience evaluation analysis is only performed on n_{17}, n_9, n_6, and n_5 instead of all gray nodes. This can effectively reduce the required time of the proposed approach.

6.3.3 Regression Analysis

In order to understand whether or not the extracted target statements from the previous phase are resilient, the original statements need to be replaced with the models that have an error. However, finding the starting point, in this case, is crucial as it may happen that the injected error to the original statement is far from the error rate that it can tolerate w.r.t the design QoS. Consequently, the target statements are incorrectly marked as non-approximable (i.e., the case of false negative). Hence, the starting error rate for the resilience evaluation of a target statement *TStm* must be derived from an alternative model that has a negligible error rate in comparison to its true behavior.

Since the target statements of our analysis approach are the computational statements, regression analysis is performed to find the starting models. In general,

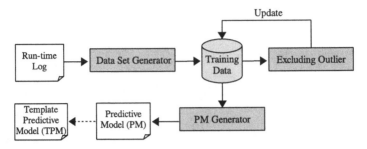

Fig. 6.3 Regression analysis overview

regression analysis is a fast machine learning technique that provides an accurate function prediction. Moreover, the estimated model obtained using the regression analysis is simple, and most importantly, has an easy interpretation. Thus it can be simply translated into the SystemC code. The main idea behind the regression analysis is to investigate the relationship between a *dependent* (response) and one or more *independent* (predictor) variables. The relationship is extracted from a finite set of given data called the training data set.

Thus, as illustrated in Fig. 6.3, the first step to perform the regression analysis is to translate the extracted information (*Run-time Log*) into the training data set (*Training Data*). Once the training data set is created, the next step is to find the closet predictive model.

6.3.3.1   Training Data Set Generation

A training data set includes two main elements which are

1. one or more inputs (the *predictors*), and
2. the corresponding output (the *response*).

Each pair of predictors and response is considered as one *observation*.

In order to translate the run-time information related to each *TStm* into a training data set, all values of its operands and result variable must be collected and isolated from others. The reason is that the run-time information is extracted at different timing points during the simulation run of the design and thus, scattered all over the *Run-time Log*. In the next step, a mapping of the *TStm* data onto the training data set is performed where the set of input operands I of the *TStm* is defined as predictors set while its output result R, as the response. This process is performed by the *Data Set Generator* module (Fig. 6.3). The following definition is used to identify this translation w.r.t Definition 6.1 formally.

Definition 6.3 For a given SystemC design including n_{TStm} target statements the training data set

$$T = \{t_i \mid t_i = \{(I, R)\} ; \ 1 \leq i \leq n_{TStm}\}$$

where each training data set t is a $(n_{op} + 1)$-dimensional vector, and the size of T is equal to the number of *TStm* specified by n_{TStm}. The parameter n_{op} identifies the number of input operands of a target statement.

In order to increase the chance of locating a proper model, the training data set is modified by excluding outliers from it. An outlier is an observation point in the training data set that is very different from other observations. The outlier is often calculated using the distance metric on the value of the response variable [53]. An observation in the training data set is considered as an outlier if the value of its response variable is more than K interquartile ranges below the lower quartile q_l or above the upper quartile q_u of its response value set. Therefore, in a training data set, all observations that their response values are out of range $[q_l - K(q_u - q_l), q_u + K(q_u - q_l)]$ are outliers. We consider the default value $K = 1.5$, which is a standard value [53].

6.3.3.2 Predictive Model Creation

A predictive model is created by determining relations between one or more predictors and the corresponding response in the training data set. The core question is whether or not a predictive model can be found to approximate the behavior of a *TStm* w.r.t the QoS of the design. The problem for a given design is formulated as the following: For each member t_i of the training data set T (based on Definition 6.3), a function $f_i : (I_i \to R_i)$ where $1 \leq i \leq n_{TStm}$, and the distribution of the predictors I_i and the function f are both unknown. The task is then to find a predictive model PM that describes the underlying data with $PM_i(I_i) \approx f_i$ for all observations in t_i with respect to the pre-defined QoS of the design.

The *PM* can be created with different regression methods depending on the observation's properties and the requirements of the resulting model. Since the distribution of predictors and response is unknown, a multiple linear regression model is applied to the training data set to find the closest model.

Algorithm 6.1 (lines 1 to 9) illustrates the process of creating the predictive model for each training data set t. The linear regression analysis is performed using three different underlying functions as the following w.r.t Definition 6.1:

- *Linear*: which only consist of linear terms in the predictors.

$$f(I) = \alpha + \Sigma_{i=1}^{n_{op}} \beta_i * op_i \tag{6.3}$$

- *Interactive*: which assumes that input operands may not be completely independent.

$$f(I) = \alpha + \Sigma_{i=1}^{n_{op}} \beta_i * op_i + \Sigma_{i=1, j=i+1}^{n_{op}-1} \beta_{ij} * op_i * op_j \tag{6.4}$$

Algorithm 6.1: Finding predictive model

Input: Training data set T
Output: Predictive model PM and QoS_{PM}
1 **foreach** *training data set* $t \in T$ **do**
2     $L_{PM} \leftarrow \emptyset$;
3     $L_{QoS} \leftarrow \emptyset$;
4   **foreach** *formula* $f \in$ *[Linear, Interactions, Quadratic]* **do**
5        $PM_f \leftarrow Regression(t, f)$;
6        $L_{PM}.add(PM_f) \leftarrow PM_f.predictorFnc()$;
7        $L_{QoS}.add(PM_f) \leftarrow PM_f.QoS$
8     $QoS_{PM}[t] \leftarrow min(L_{QoS})$;
9     $PM[t] \leftarrow L_{PM}(index(QoS_{PM}))$;

- *Quadratic*: which consists of linear, interactive, and quadratic terms in the predictors.

$$f(I) = \alpha + \Sigma_{i=1}^{n_{op}} \beta_i * op_i + \Sigma_{i=1,j=i+1}^{n_{op}-1} \beta_{ij} * op_i * op_j + \Sigma_{i=1}^{n_{op}} \beta_i' * op_i^2 \qquad (6.5)$$

where for all the aforementioned formulas, α represents the intercept, and β and β' coefficients. The parameter n_{op} is the number of predictors (operands) in the training data set t (the input operands of set I). The goal of the learning process is to estimate the intercept and coefficients of each term in order to minimize the *Root-Mean-Square Error* (RMSE) between the predicted and true models.

The result of this analysis is a predictive model PM for each training data set t that has the best QoS (i.e., RMSE) among all regression analysis. In the case of identical QoS results, the simplest model is selected to be replaced with the computational part of the design.

For example, consider the target statement S_{32} in the motivating example (Fig. 3.2, line 32). Assume that we want to estimate the behavior of the statement using the linear regression analysis. To do this, first, the training data set related to the statement is generated by analyzing its simulation behavior. In the generated training data set, the *out2* signal is defined as the response while variables *temp1* and *temp2*, as predictors. Then, the underlying formulas in (6.3), (6.4), and (6.5) are applied to the training data set to estimate the behavior of the statement. Among the aforementioned underlying formulas, the PM generated using (6.4) is the simplest model that has the lowest error. The error rate is calculated based on the RMSE between the PM and true model. In this regard, the generated PM is as the following:

$$PM = (-5.8116e - 4) + (-5.8116e - 4 * temp1) + (3.559e - 5 * temp2)$$

$$+ (temp1 * temp2) \qquad (6.6)$$

where $PM \approx out2$ and the RMSE of the PM is equal to $9.3e - 6$.

6.3.4 Statement Resilience Evaluation

The goal of this section is to find out how many of the extracted target statements of a given SystemC design can be approximated. In order to know whether or not a target statement *TStm* is resilient, the statement needs to be replaced with its corresponding generated *PM*. If the replaced model satisfies the QoS of the design, the statement is marked as approximable.

To evaluate the resilience of the target statements of the design, a *control block* is generated for each *TStm* of the design. The *control block* is an *if-else* statement that activates either the original behavior of the target statement or its approximated model. Thus, for a given target statement S_i, the *control block* is defined as the following:

$$\begin{cases} \text{if } (ctl\_S_i == true) & TPM \\ \text{else} & TStm \end{cases} \tag{6.7}$$

where ctl_S_i, *TPM*, and *TStm* indicate the control variable (which is tagged with the name of the statement), the template predictive model of the S_i statement, and the original model of the S_i statement, respectively. As illustrated in (6.7), when the control variable ctl_S_i is $true$, the estimated model *TPM* of the target statements S_i is active, otherwise, the original statement *TStm*. To evaluate the resilience of a given statement, the controllable block enables us to only activate the *TPM* of the statement under evaluation and deactivate the TPM of other target statements. Thus, once the executable model of the design is generated, it can be programmed with different controlling inputs to evaluate the resilience of each target statement.

As illustrated in Fig. 6.4, the statement resilience analysis is performed in two main phases. The first phase is to translate the generated *TPM* of each target statement of the design into the SystemC code and then replace it with the original statement. To do this, a static analysis is performed on the AST of the design's

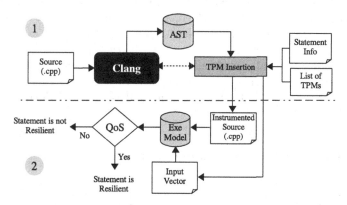

Fig. 6.4 Statement resilience evaluation overview

```
1   struct M1: sc_module{
2     ...
3     double a_S32, b_1_S32, b_2_S32, b_3_S32;   // TPM parameters
4     bool ctl_S32;  // control variable
5     ...};
6   Void M1:func1 (){
7     ...
8     if (ctl_S32 == true)
9       out1.write((a_S32) + (b_1_S32* temp1) + (b_2_S32*temp2) + (
              b_3_S32* temp1*temp2));
10    else
11      out1.write(temp2 * temp1);
12    ...}
```

Fig. 6.5 A part of the instrumented source code of the motivating example (phase 3, resilience evaluation)

source code by the *TPM Insertion* module (Fig. 6.4, phase 1) to generate an instrumented model of the design including the *control block* and the *TPM* for each target statement. To translate a *TPM* into SystemC code, the intercept and coefficients of the *TPM* are defined as global variables of the module to which the corresponding target statement belongs. The control variable of the *control block* is also defined as a global variable in the module. For example, Fig. 6.5 shows a part of the generated instrumented source code of the motivating example, including the *control block* of S_{32} (lines 8 to 11). In this figure, lines 3 and 4 show the definitions of the *TPM* parameters and control variable, respectively.

In the second phase, to determine which target statements of the design are approximable, a set of input vectors is generated by *TPM Insertion* module (Fig. 6.4) to program the *Exe Model* of the design. For a target statement *TStm*, the *Input Vector* file includes the parameters (i.e., intercept and coefficients values) of the statement's selected *PM* and the controlling signals to only activate the *PM* of *TStm* during the execution time. As shown in Algorithm 6.2 (lines 4 to 13), this process is performed for all target statements L_{TSmt} of the design. In each iteration, the executable model of the design *Exe* is executed with an input vector *IV* related to one statement in L_{TSmt} (line 8). The *IV* vector is used to program the *Exe* such that only the *PM* of the statement under evaluation is activated during the execution (line 6). As the parameters of the template predictive model in L_{TPM} is initialized by the parameters of the selected predictive models in L_{PM} (line 5), the original statement is replaced with the closest approximated model. Thus, if the QoS of the design is not satisfied (lines 9 to 12), the statement is marked as non-approximable.

6.3.5 Approximation Degree Evaluation

So far, designers are able to find the approximable parts of a given design. However, the other core questions, which have still remained, are as the following:

Algorithm 6.2: Finding approximable statements

Input: List of predictive model parameters L_{PM}, List of target statement L_{TSmt}, Design's
      executable model *Exe* and Design's *QoS*
Output: List of approximable statments L_{AS}

1 List of control variables $L_{Ctr} \leftarrow false$;
2 List of template predictive model (TPM) parameters $L_{TPM} \leftarrow \emptyset$;
3 Input Vector $IV \leftarrow (L_{Ctr}, L_{TPM})$;
4 **foreach** *statment S_i in L_{TSmt}* **do**
5      $L_{TPM}[i] \leftarrow L_{PM}[i]$;
6      $L_{Ctr}[i] \leftarrow true$;   ▷Only active the *PM* of S_i
7      $IV \leftarrow (L_{Ctr}[i], L_{TPM}[i])$;
8   run (Exe, IV);
9   **if** *QoS is satisfied* **then**
10         $L_{AS}[i] \leftarrow true$;
11   **else**
12         $L_{AS}[i] \leftarrow false$;
13      $L_{Ctr}[i] \leftarrow false$;

- Does an approximable statement accept more error while still satisfying the QoS of the design? In the case of a positive answer, what is the maximum error? (i.e., approximation degree of an approximable statement)
- How many statements of a module can be approximated at the same time? (i.e., approximation degree of a module)
- How many modules of the design can be approximated at the same time? (i.e., approximation degree of the design)

Hence, the goal of this section is to answer the aforementioned questions in order to help designers to make better decisions in approximating a given SystemC design.

6.3.5.1 Statement Approximation Degree

In order to determine the approximation degree (i.e., maximum error rate) of a statement, further analysis is applied to the approximable statements. To do this, for each statement a variant model of the statement's selected *PM* is generated (based on its *TPM*) that has more error than *PM*. To create the variant model, first, an error is injected into the response of the statement's training data set. Then, a regression analysis is applied to the new training data set based on the TPM underlying regression formula. Therefore, the generated variant model has the same underlying formula as the *PM* but with different parameters (i.e., intercept and coefficients values). The same process as in Fig. 6.4 is performed with the difference that the input vector is generated based on the new parameters w.r.t the variant model. If the QoS of design is satisfied, the error increases and the process is repeated until a variant model is generated that replacing the original statement with it causes the design's QoS violation.

Algorithm 6.3: Finding approximation degree of approximable statements

---

Input: List of template predictive model formula L_{TPM}, List of approximable target
   statement L_{ATSmt}, Design's executable model Exe, Design's QoS, Training data set
   T

Output: List of statments approximation degree L_{SAD}

1 List of control variables $L_{Ctr} \leftarrow false$;
2 List of template predictive model (TPM) parameters $L_{TPM} \leftarrow \emptyset$;
3 Input Vector $IV \leftarrow (L_{Ctr}, L_{TPM})$;
4 Error Percent $E \leftarrow -0.05 * Stm_{QoSPM}$;
5 Lower bound statement QoS threshold $Stm_{QoSL} \leftarrow \emptyset$;
6 Upper bound statement QoS threshold $Stm_{QoSU} \leftarrow \emptyset$;
7 Template Predictive Model Parameters $TPM_{paramsL}, TPM_{paramsU} \leftarrow \emptyset$;
8 **Function** $Create\_Variant (E, Bound, i)$:
9   **if** $Bound == true$ **then**
10         $T_{new} \leftarrow Lower\_Bound\_Erorr(T[i], E)$;
11   **else**
12         $T_{new} \leftarrow Upper\_Bound\_Erorr(T[i], E)$;
13      $TPM_{params} \leftarrow$ Regression $(T_{new}, L_{TPM}[i])$;
14 **foreach** $statement\ S_i\ in\ L_{ATSmt}$ **do**
15      $L_{Ctr}[i] \leftarrow true$;
16   **while** $QoS\ is\ satisfied$ **do**   ▷finding the lower bound error ot S_i
17         $Stm_{QoSL} \leftarrow E$;
18         $Bound \leftarrow true$;
19         $E \leftarrow E * 2$;
20   Create_Variant $(E, Bound, i)$;
21         $IV \leftarrow (L_{Ctr}[i], TPM_{paramsL})$;
22   run (Exe, IV);
23      $E \leftarrow 0.05 * Stm_{QoSPM}$;   ▷reset the error percent to obtain the upper bound approximation degree
24   **while** $QoS\ is\ satisfied$ **do**   ▷finding the upper bound error of S_i
25         $Stm_{QoSU} \leftarrow E$;
26         $Bound \leftarrow false$;
27         $E \leftarrow E * 2$;
28   Create_Variant $(E, Bound, i)$;
29         $IV \leftarrow (L_{Ctr}[i], TPM_{paramsL})$;
30   run (Exe, IV);
31      $E \leftarrow -0.05 * Stm_{QoSPM}$;   ▷reset the error percent for the next statement
32      $L_{SAD}[i] \leftarrow (Stm_{QoSL}, Stm_{QoSU})$;
33      $L_{Ctr}[i] \leftarrow false$;

As illustrated in Algorithm 6.3, an iterative process is performed for each approximable statement S_i in L_{ATSmt} (lines 14 to 33). In each iteration, only the corresponding control variable of the statement under evaluation is activated (line 15). For statement S_i, two degrees of approximation are calculated based on the lower (lines 16 to 22) and upper (lines 24 to 30) bounds error that the statement can accept while satisfying the QoS of the design. The error rate (for both the lower and upper bounds) is calculated by the *Creat_Variant* function (lines 8 to 13). It is based on injecting error E into the training data set T (lines 10 and 12) of the statement and applying the regression analysis (line 13) to T using the *TPM* underlying formula of the statement. To find the lower and upper bounds approximation degree, the initial value of E is set to -5% (line 4) and $+5\%$ (line 23) of the error between the original statement and its selected *PM* (Stm_{QoSPM}), respectively. The error is injected into the response values in the training data set of the statement. Each time the design's QoS is satisfied, the error rate E two times increases (lines 19 and 27). This process is repeated until the design's QoS is violated. The error rate before this violation is considered as the approximation degree of the statement (lines 17, 25, and 32) which are Stm_{QoSU} (upper bound approximation degree) and Stm_{QoSL} (lower bound approximation degree).

For example, consider the target statement S_{32} in the motivating example (Fig. 3.2, line 32) where the goal is to find its upper bound approximation degree Stm_{QoSU}. Assume that the QoS of the design is given based on the RMSE for the output signal *Out2* of module *M2*, which is equal to $2.53e - 4$. To find the Stm_{QoSU} based on the Algorithm 6.3, the initial value E is set to 5% of its *PM* error rate ($9.3e - 6$) w.r.t (6.6) which is equal to $4.65e - 7$. After ten iterations, a QoS violation is reported. It means that until nine iterations, the QoS of the design is satisfied thus the Stm_{QoSU} is equal to $2.38e - 4$.

The number of iterations to find the approximation degree of a statement is equal to $\log(Stmt_{QoSU} * Stmt_{QoSL}/Stmt_{QoSPM})$ in the worst case. Therefore, it has a logarithmic time complexity.

6.3.5.2 Module Approximation Degree

For approximating a given SystemC design, designers may be interested in approximating a portion of the design at a higher granularity level than a statement (e.g., a module). In this case, knowing the *Module Approximation Degree* (MAD) helps them to make a better decision. The approximation degree of a module provides designers with the maximum number of statements that can be approximated at the same time in the module w.r.t QoS of the design. Based on the number of approximable statements in a module, the maximum error that the module can accept (which is propagated to its final output) is also identified.

To find the approximation degree of a module, several corner cases of error models can be considered. This can be done by adjusting different error rates to each approximable statement (i.e., the minimum error rate that the statement can accept to the maximum error rate) and then by creating different statements' combinations from the whole approximable statements of the module. However, this leads to a very time-consuming task even if a fully automated analysis is performed. Assume that n target statements of a module are marked as approximable and on average m variants are required to obtain the maximum error rate of each statement. To cover all corner cases in the approximation degree analysis of the module, $(m^n - n - 1)$ simulation runs are required. As a simple estimation of the required time for this analysis, consider a module that has ten target statements marked as approximable. On average, to find the approximation degree of each statement, ten simulation runs were performed. It means that ten variants (with different errors) of the statement were generated and the statement approximation degree (maximum error) was obtained in the tenth simulation run. Assume that each simulation run requires 0.01 s. In this respect, the overall required time of this analysis is about 3 years.

Hence, we only consider two important corner cases of error models where different combinations of the approximable statements of a module for the best (Stm_{QoSPM}) and worst (Stm_{QoSL} and Stm_{QoSU}) cases of error related to the statements are evaluated. The former error model gives designers an estimation of the maximum number of statements that can be approximated at the same time with their minimum error (based on the statement selected PM error i.e. Stm_{QoSPM}) in a module. This is useful when designers want to apply a $Mild$ approximation to the design.

The latter error model provides designers with an estimation of the maximum number of statements that can be approximated at the same time with their maximum error (based on their lower (Stm_{QoSL}) and upper (Stm_{QoSU}) error bounds) in a module. This is useful when designers want to apply an $Aggressive$ approximation to the design.

The process of finding MAD in the case of $Mild$ approximation is illustrated in Algorithm 6.4 (lines 1 to 11). It includes an iterative loop (lines 5 to 11) wherein each iteration, a combination of the approximable statements in L_{ATSmt} is selected, and the corresponding input vector IV is generated (lines 6 and 7). The executable model of the design Exe is run with the generated input vector (line 8). If the QoS of the design is satisfied, the statement combination SC_i is stored in MAD_{mild} (line 10) and iteration is stopped (line 11). In order to reduce the required time of the analysis, iterations are started from the combinations with the maximum number of statements (i.e., the best case, all statements can be approximated at the same time) to the minimum number (i.e., the worst case, two statements can be approximated at the same time). Moreover, we take advantage of the $Statement\ Info$ generated in the first phase of the proposed approach (e.g., Fig. 6.2) to reduce the number of statements' combinations. In this regard, the approximable statements which are in the same group w.r.t (6.2) are approximated at the same time. The reason is that different statements in the body of a conditional statement cannot be executed simultaneously. At a specific time, only one of these statements is executed.

For example, consider the motivating example (Fig. 3.2) where the goal is to find the $Mild$ approximation degree of module $M1$. Assume that the QoS of the design is given for the output signal $Out2$ of module $M2$. Based on Algorithm 6.4, L_{ATSmt} includes target statements S_{24} ($temp1$, line 24), S_{27} ($temp1$, line 27) and S_{29} ($temp2$, line 29), S_{32} ($out2$, line 32). Due to the $Statement\ Info$, statements S_{24}, S_{27} are in the same group. To find the module degree, the analysis is started by evaluating the statement combinations $SC_1 \leftarrow (\{S_{24}, S_{27}\}, S_{29}, S_{32})$ which has the maximum number of statements. If the QoS of the design is satisfied, SC_1 with four statements is stored in MAD_{mild}; otherwise, the other three possible combinations,

---

Algorithm 6.4: Finding the approximation degree of a module

Input: List of template predictive model formula L_{TPM}, List of approximable target
statement L_{ATSmt}, Design's executable model Exe, Design's QoS

Output: module approximation degree MAD_{mild} and MAD_{agg}

1   List of control variables $L_{Ctr} \leftarrow false$;
2   List of template predictive model (TPM) parameters $L_{TPM} \leftarrow \emptyset$;
3   Input Vector $IV \leftarrow (L_{Ctr}[\emptyset], L_{TPM}[\emptyset])$;
4   Template predictive model parameters $TPM_{params} \leftarrow \emptyset$;
5   **foreach** *statement combination SC_i in L_{ATSmt}* **do**
6       $L_{Ctr}[i] \leftarrow true$;
7       $IV \leftarrow (L_{Ctr}[i], TPM_{params}[i])$;
8   run (Exe, IV);
9   **if** *QoS is satisfied* **then**
10         $MAD_{mild} \leftarrow SC_i$;
11   break;

12   List of abandoned combinations $L_{abandoned} \leftarrow \emptyset$;
13   **foreach** *statment combination SC_i in L_{ATSmt}* **do**   ▷for all unique two-statement combinations
14       $flag \leftarrow EV(SC_i)$;
15   **if** *flag* **then**
16   break;

17   **Function** $EV (SC_i)$:
18       $L_{Ctr}[i] \leftarrow true$;
19       $IV \leftarrow (L_{Ctr}[i], TPM_{params}[i])$;
20   run (Exe, IV);
21   **if** *(QoS is satisfied)* **then**
22   **if** *length (SC_i) > length of exsiting combinations* **then**
23           $MAD_{agg} \leftarrow SC_i$;
24   **return** true;
25   **else**
26           $SC_i \leftarrow AddStm(SC_i, L_{ATSmt})$; ▷creating a new combination by adding a statement
27   **if** *SC_i not $\cap$ $L_{abandoned}$* **then**
28             $EV (SC_i)$;

29   **else**
30         $L_{abandoned} \leftarrow SC_i$; ▷abandoning the combination that causes QoS violation
31   **return** false;

---

i.e., $SC_2 \leftarrow (\{S_{24}, S_{27}\}, S_{29})$, $SC_3 \leftarrow (\{S_{24}, S_{27}\}, S_{32})$, and $SC_4 \leftarrow (S_{29}, S_{32})$ are evaluated.

In the worst case, the required number of simulation runs for this analysis is $(2^{n_{AStm}} - n_{AStm} - 1)$, where n_{AStm} is the number of approximable statements in L_{ATSmt}. Thus, the time complexity of the algorithm is an order of $2^{n_{AStm}}$ in the worst case. Due to the number of approximable statements of a module, the required time of this analysis might be large in the worst case. Hence, we define the parameter $n_{threshold}$ (as an interactive option for designers) to restrict the analysis to be performed for a maximum of $2^{n_{threshold}}$ iterations. This interactive option provides designers with a trade-off between the accuracy of the analysis and its required time.

Regarding MAD in the case of *Aggressive* approximation, as each approximable statement has two error models (Stm_{QoSU} and Stm_{QoSL}), MAD needs to be identified w.r.t each of them. Hence, MAD_{aggU} and MAD_{aggL} refer to the maximum number of statements that can be approximated at the same time in a module with their Stm_{QoSU} and Stm_{QoSL} error models, respectively.

As an instance, to find MAD_{aggU}, we do not consider all possible combinations of approximable statements. We take advantage of the data dependency graph to limit the number of statement combinations. The reason is that approximating two dependent statements where both are approximated with their upper bound error, Stm_{QoSU} tends to an accumulative error (suppose ΔE) which is propagated to the final output. If the generated error (ΔE) is more than acceptable QoS of the design, all statement combinations including these two statements can be eliminated from the state space of statement combinations (as they include at least ΔE).

For example, consider the data dependency graph (Fig. 6.2) of the motivating example (Fig. 3.2). Assume that the goal is to find the MAD_{aggU} of module *M1* where the QoS of the design is given for node n_{19} (output signal *Out2* of module *M2*). Analyzing the data dependency graph shows that only nodes n_5, n_6, and n_8 of module *M1* are in the cone of n_{19} (i.e., contribute to generating the final result). To find the MAD_{aggU}, the statement combinations are started from a unique combination of two statements, e.g., n_5 and n_8. Due to the accumulative error, if the statement combination, including nodes n_5 and n_8 causes a QoS violation, all other statement combinations including these two nodes also cause a QoS violation. Thus, the statement combinations including n_5, n_6, and n_8 are eliminated from the state space of statement combinations. In contrast, if the QoS of the design is satisfied, one statement is added to the existing combination, and the new statement combination is evaluated. This process continues until either no statement combination remains or the current statement combination has the maximum number of statements as other larger combinations have been eliminated earlier.

The process of finding MAD in the case of *Aggressive* approximation is illustrated in Algorithm 6.4 (lines 12 to 31). First, a statement combination SC_i, including two statements, is selected and evaluated using *EV* function (lines 13 and 14). The *EV* function generates the corresponding input vectors (lines 18 and 19) and runs the executable model with the generated vector (line 20). If the design's QoS is satisfied (lines 21 to 28), it checks that the current combination SC_i has the greatest number of statements among the existing statement combinations. If true, SC_i is stored in MAD_{agg} and the process ends (lines 22 to 24). Otherwise, a new statement combination is generated by adding a new statement to the existing SC_i that is not an abandoned combination (line 26 and 27). Then, the *EV* function is called to evaluate the new SC_i (line 28). If the design's QoS is not satisfied (lines 29 to 31), SC_i is abandoned and the *EV* function returns false. It means that the next statement combination (including two statements) is selected and the process is repeated (line 13).

As the process of finding MAD in the case of *Aggressive* approximation is based on a *Depth-First Search* (DFS) on the data dependency graph for all statement combinations with length two, it comes with linear complexity in the worst case.

6.3.5.3 Design Approximation Degree

In order to find the maximum number of modules that can be approximated simultaneously, a similar analysis to Algorithm 6.4 is applied to the whole approximable statements of the design. The only difference is that the statement combinations must at least include two statements from two different modules.

In the case of *Mild* approximation (w.r.t Algorithm 6.4), the number of simulation runs is $(2^{n_{DAStm}} - n_{module} - 1)$ in the worst case, where n_{DAStm} and n_{module} represent the number of approximable statements of design and the number of modules, respectively. Therefore, the time complexity of this analysis is the same as Algorithm 6.4 (lines 1 to 11). However, due to the number of design's statements (which is greater than the number of module statements), this analysis may require significant time in the worst case. Hence, the same as MAD, we define the parameter $Stm_{threshold}$ (as an interactive option for designers) as the maximum number of statements to restrict the analysis to be performed only when n_{DAStm} is equal or less than $Stm_{threshold}$. For a design with the number of approximable statements more than $n_{threshold}$, we only consider the combination of modules with their maximum degree of approximation for the three error models Stm_{QoSPM}, Stm_{QoSL}, and Stm_{QoSU}.

Regarding *Aggressive* approximation, the complexity of this analysis (w.r.t Algorithm 6.4, lines 12 to 31) is linear.

6.4 Experimental Evaluation

The proposed approach was applied to several standard ESL benchmarks provided by [97] and OSCI [1].

First, a case study from image processing domain is illustrated in detail in Sect. 6.4.1. Second, we give a brief discussion on the obtained experimental results to evaluate the quality of our approach in Sect. 6.4.2. To quantify the QoS loss of each design due to approximation, we consider *Peak Signal-to-Noise Ratio* (PSNR) for the *JPEG-encoder* and *Sobel* designs and RMSE for all other designs. All experiments were carried out on a PC equipped with 8 GB RAM and an Intel Core i7 CPU running at 2.4 GHz.

6.4.1 Case Studies

The experimental results for different types of ESL benchmarks are illustrated in Tables 6.1 and 6.2. In both tables, the first three columns (*ESL Model*, *LoC*, and *#M*) list the name, lines of code, and number of modules of each design, respectively.

In Table 6.1, column *#AM* shows the number of modules that were detected as approximable by the proposed approach. Column *AMN* represents the name of the approximable modules detected by the proposed approach. Columns *#F* and *AFN* show for each approximable module, the number of its functions and the name of the functions which include at least an approximable statement, respectively. Column *#Obsv* lists the number of observations used in the regression analysis to estimate the behavior of each target statement. The *Outl* column shows the percentage of outliers in the observation data that was excluded from the training data set. These values are obtained based on the formula in Sect. 6.3.3.1. The *#MTStm* column represents the number of target statements of each module. Column *#ATStm* illustrates the number of target statements of each module marked as approximable.

Column *MAD* shows the approximation degree of each module based on the maximum number of its approximable statements that can be substituted with the corresponding original statements at the same time. The approximation degree is reported for the *Mild* (column *Mld*) and *Aggressive* (column *Agg*) approximations of the design. In the case of *Aggressive* approximation, the maximum number of approximable statements between the upper (Stm_{QoSU}) and lower (Stm_{QoSl}) error bounds is reported. As illustrated in this column, the approximation degree of the *exec* module of the *RISC-CPU* design is 13 for both *Mild* and *Aggressive* approximation models. The reason is that all approximable statements of the modules belong to a conditional block (a *switch-case* statement). Thus, they can be approximated at the same time as during the execution, at a specific time, only one of them is executed.

Column *DAD* represents the approximation degree of the entire design. It indicates the maximum number of modules (with the maximum number of approximable statements) that can be approximated at the same time. The approximation degree is reported for the *Mild* (column *Mld*) and *Aggressive* approximations (column *Agg*) of the design. In the case of *Aggressive* approximation, the maximum number of the approximable module between the upper (Stm_{QoSU}) and lower (Stm_{QoSl}) error bounds is reported.

Table 6.2 demonstrates the required time of the proposed approach for resilience evaluation analysis of the SystemC designs. Column *ET* illustrates the execution time of the proposed approach based on the required time for each phase. The *Phase1*, *Phase2*, and *Phase3* columns show the execution time of the information extraction, regression analysis, and statement resilience evaluation phases of the proposed approach. Column *Phase4* illustrates the execution time of the approximation degree evaluation phase of our approach, followed by the required time to obtain the approximation degree of each approximable statement (column *StmAD*), each module (column *MAD*), and the entire design (column *DAD*). For designs with

Table 6.1 Experimental results related to the quality of the proposed resilience evaluation approach for all case studies

ESL model	LoC	#M	#AM	AMN	#F	AFN	#Obsv	Outl%	#TStm	#MTStm	#ATStm	MAD Mld	MAD Agg*	DAD Mld	DAD Agg*
Cholesky[2]	489	1	1	Cholesky	2	chlky_cmp	250	0	3	3	3	2	2	1	1
3-stage pipe[1]	511	5	3	Stage1	1	Addsub	500	0	5	2	2	2	2	2	2
				Stage2	1	Multdiv	500	0.0260		2	2	2	2		
				Stage3	1	pwr	500	0		1	1	1	1		
Fft-flp[1]	586	3	1	fft	1	Entry	500	0.0100	15	15	13	6	5	1	1
Interpolation[2]	587	1	1	Interp	1	Run	500	0.0120	5	5	5	3	2	1	1
Sobel[2]	713	1	1	Sobel	2	sobel_filter	131,072	0.0020	3	3	3	2	1	1	1
IDCT[1]	809	1	1	idct	2	idct_islow	500	0.0120	20	20	14	8	6	1	1
FIR[1]	834	2	2	fir	1	Entry	500	0.0100	19	3	3	2	1	2	1
				fir_data	1	Entry	500	0.0100		16	14	8	4		
Decimation[2]	893	1	1	decfilt	1	Run	500	0.0120	10	10	5	3	2	1	1
RISC-CPU[1]	1960	3	3	exec	1	Entry	1250	0.0520	61	13	13	13	13	3	3
				Floating	1	Entry	1250	0.0504		15	10	8	6		
				mmxu	1	Entry	1250	0		27	15	11	8		
JPEG-encoder[2]	1422	5	4	dct	1	jpeg_dct	131,072	0	9	1	1	1	1	4	3
				Huffman	4	jpeg_AC	131,072	0		3	3	4	2		
						jpeg_DC	131,072	0		2	1				
				Quantization	1	jpeg_quant	131,072	0		2	1	1	1		
				rle	1	jpeg_rle	131,072	0		1	1	1	1		

[1]OSCI and [2][97]

LoC lines of code, *#M* number of modules, *#AM* number of approximable modules, *AMN* approximable module name, *#F* number of functions, *AFN* approximable function name, *#Obsv* number of observations, *Outl%* percentage of outlier, *#TStm* number of target statements execution, *#MTStm* number of module target statements, *#ATStm* number of approximable target statements, *MAD* module approximation degree, *DAD* design approximation degree

*The reported number is based on the maximum degree between upper and lower bounds error

148 6 Application III: Design Space Exploration

Table 6.2 Experimental results related to the execution time of the proposed resilience evaluation approach for all case studies

ESL Model	LoC	#M	ET (s)			Phase4				CET (s)		
			Phase1	Phase2	Phase3	StmAD	MAD	DAD	Total	Compile	Exec	Total
Cholesky[2]	489	1	5.19	1.31	5.87	0.36	0.15	0	12.88	4.62	0.003	4.63
3-stage pipe[1]	511	5	4.77	1.38	5.19	0.45	0.28	0.36	12.43	4.12	0.010	4.13
Fft-flp[1]	586	3	5.41	2.01	5.80	2.54	31.52	0	47.28	3.94	0.004	3.95
Interpolation[2]	587	1	5.94	2.12	6.07	0.55	0.48	0	15.16	4.79	0.007	4.86
Sobel[2]	713	1	3.02	1.18	3.29	0.36	0.15	0	12.88	2.15	0.231	2.38
IDCT[1]	809	1	5.13	2.81	5.71	69	109.62	0	132.96	3.51	0.019	3.53
FIR[1]	834	4	5.29	2.70	5.90	10.22	63.61	3.71	91.43	3.82	0.010	3.83
Decimation[2]	893	1	5.59	1.27	6.18	2.83	4.64	0	20.51	4.69	0.021	4.90
RISC-CPU[1]	1960	10	13.32	3.51	14.02	11.22	29.13	0.01	71.21	10.08	0.011	10.19
JPEG-encoder[2]	1422	5	8.79	11.41	9.36	793.92	69.92	167.52	1060.92	7.26	9.561	16.82

[1]OSCI and [2][97]

LoC lines of code, *#M* number of modules, *ET* execution time in seconds, *CET* compilation and execution time in seconds, *StmAD* statement approximable degree statements, *MAD* module approximation degree, *DAD* design approximation degree

only one approximable module, the reported time in column *DAD* is zero as no modules combinations need to be performed. In the case of *RISC-CPU* design, the reported time in column *DAD* is the time that was spent to analyze its data dependency graph to find out the approximable modules are independent. Thus, no module combination analysis was performed. Column *CET* shows the time of each design's compilation and execution without any instrumentation.

In the following, to evaluate the quality of the proposed approach, we consider the *JPEG-encoder* design [97] in detail, which is widely used in the image processing domain as compression algorithm.

6.4.1.1 JPEG-Encoder

The JPEG-encoder design is the hardware implementation of the encoder part of the JPEG standard (a common compression method for digital images) in the SystemC language, which is provided by Schafer and Mahapatra [97]. To quantify the QoS loss of the design due to the approximation, two values of the PSNR w.r.t the approximation models are defined. The minimum value of PSNR is assigned to 20 dB and 15 dB for the *Mild* and *Aggressive* approximation models, respectively.

As shown in Table 6.1, the design includes nine target statements (column *TStm*), while seven of them were marked as approximable. To evaluate the approximation degree (Stm_{QoSU} and Stm_{QoSL}) of each approximable statement of the design, as illustrated in Fig. 6.8, overall, 82 iterations w.r.t Algorithm 6.3 were performed.

The approximable statements belong to four modules of the design, which are *dct*, *huffman*, *quantization*, and *rle*. Take e.g. the *huffman* module of the design that four out of its five target statements were marked as approximable. The statement that was marked as non-approximable belongs to the *jpeg_addzero* function of the *huffman* module that its result variable is used in the condition statement of a control block. Figure 6.6a shows the original output image of the *JPEG-encoder* design where no approximation was performed. Figure 6.6b illustrates

(a) Original   (b) Mild approximation   (c) Aggressive approx-   (d) Aggressive approx-
             with Stm_{QoSPM} imation with Stm_{QoSU} imation with Stm_{QoSL}

Fig. 6.6 Image (**a**) is the original output of *JPEG-encoder* design. (**b**) is the output image of approximating the *huffman* module of the JPEG-encoder with its *mild* degree of approximation. Images (**c**) and (**d**) are the result of approximating the *huffman* module with its *aggressive* degree of approximation for the upper and lower error bounds, respectively

| (a) Original | (b) Mild approximation with Stm_{QoSPM} | (c) Aggressive approximation with Stm_{QoSU} | (d) Aggressive approximation with Stm_{QoSL} |

Fig. 6.7 Image (**a**) is the original output of *JPEG-encoder* design. (**b**) is the output image of approximating the *JPEG-encoder* design with its *mild* degree of approximation. Images (**c**) and (**d**) are the result of approximating the *JPEG-encoder* design with its *aggressive* degree of approximation for the upper and lower error bounds, respectively

the result of *Mild* approximation of *huffman* module where its four approximable statements (i.e., its approximation degree) were replaced with their corresponding selected PMs (which have the minimum error rates i.e. Stm_{QoSPM}) at the same time. The selected statement combination includes three target statements of the*jpeg-AC* function and one target statement of the *jpeg-DC* function. Thus, the result of this analysis was obtained in the first iteration (the best case) where a combination of all approximable statements of the module is evaluated. The output image of the *JPEG-encoder* for the *Mild* approximation of the *huffman* module has PSNR 21.93 dB in comparison to the original output image. Thus, it satisfies the QoS of the design. In the case of *Aggressive* approximation, Fig. 6.6c and d represent the design's output images where two target statements of the module were replaced with the corresponding *TPMs* which have the maximum upper (i.e., Stm_{QoSU}) and lower (i.e., Stm_{QoSL}) error bounds, respectively. The output image of the *JPEG-encoder* for the *aggressive* approximation of the *huffman* module in comparison to the original output image has PSNR 15.21 and 16.04 dB for the Stm_{QoSU} and Stm_{QoSL} error models, respectively.

As illustrated in Fig. 6.8, the required number of iterations w.r.t Algorithm 6.4 to find the approximation degree for all module of the design is seven. Since the other three modules of the design have only one approximable statement (thus no statement combinations need to be performed), all iterations were performed to find the approximation degree of the *huffman* module.

Regarding the approximation degree of the design, in the case of *Mild* approximation, the maximum number of modules that can be approximated at the same time is four. It means that there is at least one statement combination of the design's approximable statements where each module has at least one approximable statement. The first statement combination, which was found by the proposed approach, includes one statement of module *dct*, one statement of module *quantization*, and two statements of module *huffman*. The output image for the *Mild* approximation of the *JPEG-encoder* is shown in Fig. 6.7b. It has PSNR 20.24 dB in comparison to the original output image in Fig. 6.7a. Thus, it satisfies the QoS of the design.

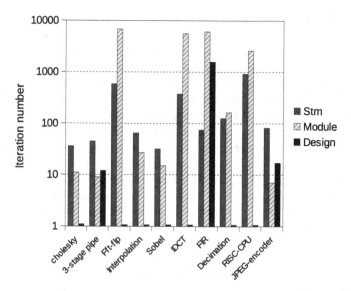

Fig. 6.8 Total number of iterations to find the approximation degree of approximable statements, modules, and design

The approximation degree of the design w.r.t the *Aggressive* approximation for both error models (i.e., the maximum upper Stm_{QoSU} and lower Stm_{QoSL} error bounds) is three. Therefore, as maximum, three modules of the design can be approximated at the same time such that each module has at least one approximable target statement with its maximum error rate. The first statement combination found by the proposed approach w.r.t Stm_{QoSU} is a combination including three approximable statements of modules *dct*, *quantization*, and *huffman*. Regarding Stm_{QoSL}, the first statement combination is a combination of four approximable statements consisting of two statements of the *huffman* module, and one statement of *dct* and *quantization* modules.

The output images for the *Aggressive* approximation of the *JPEG-encoder* related to the Stm_{QoSU} and Stm_{QoSL} error models are presented in Fig. 6.7c, d, respectively. In comparison to the original output image, the PSNR for the former model is 15.01 dB, for the latter 15.89 dB. Thus, both models meet the QoS of the design. As shown in Fig. 6.8, this analysis requires 17 iterations w.r.t Sect. 6.3.5.3.

6.4.2 Integration and Discussion

The proposed approach is able to identify the resilient portions of a given SystemC design without needing designers to have prior knowledge of the design domain (i.e., a black-box testing framework for resilience evaluation). Moreover, if designers have knowledge about a given design (e.g., modules that consume high amounts of power or have significant effect on performance), the proposed approach can be applied only to those specific modules instead of the entire design. The approximation degree analysis of the proposed approach effectively guides designers to have an estimation of a wide range of error rates that a design portion can accept at different granularity levels. This provides designers with various options to choose the best approximation technique based on their requirements, especially in the case of complex or legacy models.

The approximation degree of a resilient statement gives an estimation of the minimum (i.e., based on the $Stmt_{QoSPM}$) and maximum (i.e., based on the $Stmt_{QoSU}$ and $Stmt_{QoSL}$) error rates that the statement can accept w.r.t to the QoS of the design. The module approximation degree (MAD) provides designers with the maximum number of approximable statements that can be approximated at the same time w.r.t the design's QoS. MAD is reported for the best and the worst case error rates of the approximable statements. If the granularity level of approximation is defined based on module approximation, MAD gives designers an estimation of the error rate that the module can tolerate while meeting the QoS of the design. In this respect, the approximation degree of the design (DAD) shows the maximum number of modules that can be approximated at the same time. Thus, even in this granularity level designers have this option to choose the best alternative of approximating the entire module w.r.t to its maximum error rates.

A direct application of the proposed approach is a feedback system for designers in the synthesis process. It provides designers with the information about selecting an approximation technique (either based on modifying operands, i.e., variables of the design or operators) to approximate resilient portions while synthesizing the design to the lower levels of abstraction. As a simple example, if designers want to synthesis the design to the lower levels of abstraction (e.g., the RTL or below), the approximation degree of a resilient statement can help them to choose an approximable component from the existing library (e.g., an adder) which has the error rates (lower and upper bounds) less or equal to the statement approximation degree.

The required time of the proposed approach to analyze a given SystemC design is divided into two main parts. First, the time that is spent to identify resilient portions of the design including the run-time information extraction (phase 1), the regression analysis on the extracted information (phase 2), and evaluating whether or not a given portion is resilient or not (phase 3). As illustrated in Table 6.2, the execution time of each of the aforementioned phases is in order of seconds. In comparison to the pure compilation time of each design (column *compile*), these time frames are in a reasonable boundary. The complexity of the regression analysis (phase 2) depends on the creation of the training data sets and finding the predictive model. As we take advantage of the linear regression (which is a fast machine learning technique with high accuracy for function estimation), the complexity of this analysis is linear.

The second part of the required time of the proposed approach belongs to the approximation degree analysis (phase 4). This time relies on the number of approximable statements of the design. The major time-consuming part of this phase is the approximation degree analysis for *Mild* approximation model where the time complexity in the worst case is exponential. However, using results of the static analysis in the first phase of the proposed approach (i.e., data dependency graph and *Statement Info*), we could reduce the number of statement combinations and consequently, the execution time of this analysis. Moreover, the proposed approach provides designers with some interactive options to control the complexity of this analysis. Overall, as illustrated in Fig. 6.8 and Table 6.1 (column *#ATStm*), the more number of approximable statements (which do not belong to the same conditional block) a design has, the more number of iterations are required for approximation degree analysis. The largest execution time reported in Table 6.1 (the *JPEG-encoder* design) is about 18 min, which is still a reasonable time frame, allowing the proposed approach to be used in common development environments.

6.5 Conclusion

In this chapter, we presented a new approach to automatically detect the approximable portions of ESL designs for the task of design space exploration. The proposed approach can effectively guide designers to know under which error limits, different portions of a given SystemC design at different granularity levels (i.e., from a single statement to a module) can be approximated. The approach is based on a combination of static and dynamic analysis methods to extract the simulation behavior of different portions (i.e., computational statements) of the design. To perform this hybrid analysis, we take advantage of the compiler-based approach (introduced in Chap. 3). We translated the extracted simulation behavior of each portion into a training data set and performed regression analysis techniques to estimate its true behavior with a predictive model. The generated model is used as the starting point of the portion resilience evaluation analysis to reduce the false-negative case of finding the resilient portions. Moreover, we performed an approximation degree analysis to determine the maximum error rate that an approximable statement can accept. Subsequently, the maximum number of statements and modules that can be approximated at the same time were reported to designers. Several ESL benchmarks were run to evaluate the effectiveness of our approach. Initial results sound promising, and we believe this new line of research is helpful for making approximate computing truly a cross-cutting activity in the early stages of the design process.

Chapter 7
Conclusion

In this book, a novel design understanding methodology was presented, enabling designers to analyze a given SystemC-based VP from two perspectives: the debugger-based and the compiler-based approaches.

The former approach provides designers with a non-intrusive analysis solution that only requires the executable model of VPs. Thus, in case of legacy or third-party IPs where the source code may not be available at all, it is the only applicable solution. It was shown how static information can be extracted from the debug symbols of a given VP and be used to program the debugger to retrieve the VP's behavior at run time.

The latter approach introduces a fast analysis technique which scales very well with an arbitrary complexity of VPs or their running software (or application). The required static information is extracted by analyzing the AST of VPs and used to generate an instrumented version of the VP for run-time information retrieval.

In comparison to the previous SystemC designs analysis methods, the main advantages of both proposed approaches are as follows:

- They require no prior limitation (or assumption) on the SystemC language. To analyze a given VP, they do not rely on any assumption about, e.g., a well-defined subset of SystemC used in the VP.
- They are fully automated and do not need any manual programming effort or annotation to the existing framework (i.e., SystemC kernel, interfaces, or library) by designers.
- They support both the SystemC cycle-accurate models and VPs modeled using TLM-2.0 constructs, while the existing methods are mostly applicable for either of them.
- They are able to extract precise information related to both the structure and the run-time behavior of SystemC-based VPs, whereas most of the existing methods extract a limited set of information that only describes the structure of the model and does not reflect its run-time behavior.

© Springer Nature Switzerland AG 2020
M. Goli, R. Drechsler, *Automated Analysis of Virtual Prototypes at the Electronic System Level*, https://doi.org/10.1007/978-3-030-44282-8_7

Apart from the aforementioned advantages, we showed how the extracted information (i.e., the structural and behavioral) using the proposed approaches could be visualized to effectively guide designers to understand the intricacy of a given VP. Moreover, the interactive options provided by the proposed approaches enable designers to only focus on specific features of VPs, resulting in gaining performance or decreasing analysis time.

The information related to the VP's structure was presented as an XML formatted file, providing designers with a structural presentation of the design architecture. Moreover, it can be easily read and parsed by future tools which may use the result of our analysis approaches.

The extracted information related to the SystemC TLM-2.0 VPs' behavior was translated into a set of UML diagrams. The UML diagrams, generated by the proposed methodology, reflect the behavioral information in a big scale view to empower designers to easily trace both transactions' data and flow at the same time. Moreover, a classification algorithm was introduced for the first time to eliminate the redundant transactions which are unnecessary from the perspective of design understanding. We illustrated how the extracted transactions are categorized based on their flow and type to be used for the visualization of those transactions presenting a unique behavior. This allows designers to quickly grasp the VP's behavior, especially in the case of complex or legacy models that may include thousands of transactions.

In the case of SystemC cycle-accurate VPs, the VPs' behavior is presented as a VCD file. In comparison to the existing counterpart (SystemC API, which is the conventional solution to monitor the run-time behavior of SystemC cycle-accurate VPs), it has four main advantages as follows:

- Tracing the base type variables (e.g., C++ data type) that may change several times during a single SystemC-δ-cycle (the intra-cycle feature of the proposed approaches).
- Extracting the user-defined data types that are not supported by SystemC API at all unless the designers alter their code.
- Tracing the values of local variables of modules and functions.
- Eliminating the manual programming effort required by SystemC API to modify the source code and include all signals that need to be traced.

Therefore, concerning the first goal of this book, a comprehensive design understanding methodology for analyzing VPs at the ESL was presented, which is able to support a wide variety of SystemC designs and can be used in conjunction with an existing framework or any SystemC setup.

Regarding the second goal of this book, various applications were implemented as use cases and examples of the design understanding results. The following applications were served to show how the extracted information could be used in practice off the shelf or with minimum translation effort during the design process.

- **Verification:** the extracted information was used to formally verify the simulation behavior of a given SystemC TLM-2.0 VP against TLM rules and the VP's specifications. In this regard, we showed how this information is translated into a set of timed automata models, and the TLM rules as well as VP's specifications into a set of properties w.r.t TCTL property language. Then, a model checker was used to validate the former against the latter. By introducing this semi-formal verification approach, we could effectively overcome the drawbacks of the previous verification methods at the ESL which are in short (1) lack of verifying VPs against their specifications, (2) state space explosion and limitation of modeling VPs in formal semantics (for formal methods), (3) lack of supporting a wide range of VPs, and (4) low degree of automation.
- **Security Validation:** as the second application, a novel automated approach was presented to validate the security of a given VP against two security threat models which are information leakage (confidentiality) and unauthorized access to data in a memory (integrity). We illustrated how the extracted information (related to the VP's transactions) and the VP security rules are automatically translated into a set of access paths and properties, respectively. Then, the translated model was validated against the security properties. The proposed security validation approach is *the first of its kind* at the ESL, which could effectively eliminate the drawbacks of the existing static security validation analysis. In case that the address of transactions is defined at run time, e.g., generated either explicitly by initiator modules (based on some dynamic computation) or implicitly by its running software, static analysis methods are not able to detect this security violation.
- **Design Space Exploration:** as the last application, the information extraction analysis was used to detect the approximable portions of ESL designs automatically. A new approach is presented based on a combination of static and dynamic analysis methods, along with regression analysis techniques. We showed how the extracted simulation behavior of each VP's portion is translated into a set of training data and used as the input of regression analysis techniques. The generated predictive model was used as the starting point of a portion resilience evaluation to reduce the false-positive case of finding the resilient portions. The proposed approach can effectively guide designers to know under which error limits different portions of a given SystemC design at different granularity levels (i.e., from a single statement to a module) can be approximated. Moreover, an approximation degree analysis was performed to determine the maximum error rate that an approximable statement can accept. Subsequently, the maximum number of statements and modules that can be approximated at the same time are reported to designers.

7.1   Outlook

The comprehensiveness and automation of the proposed approaches in this book are their main advantages that enable designers to tackle basically all aspects of SystemC-based VPs analysis. However, there are still open questions that are worth to be investigated in further research.

One possible future direction respecting the design understanding is to support VPs, including user-defined TLM protocols. Moreover, it would be very interesting to modify the proposed approach to support the design understanding of both the software and hardware parts of a given VP-based SoC at the same time. This requires to extract the information of the software running on the underlying hardware and also a comprehensive visualization model that empowers designers to trace the behavior of the entire system. In addition to the above mentioned, it would be interesting to investigate if the proposed approaches can also be applied to SystemC models implemented using the SystemC *Analog Mixed Signal* (AMS) extensions. Although our proposed approaches potentially are able to support SystemC-AMS, to extract the information and present them in a proper format, they must be modified based on the information which is useful in this domain to be extracted and presented.

Regarding the design understanding application, despite the success of the presented applications, there are still opportunities for further enhancements, e.g., supporting verification of user-defined TLM protocols, functional and timing verification of SystemC cycle-accurate VPs, or validation of other security threat models. Moreover, as the results of the design understanding phase provide designers with an intermediate representation of the whole system's behavior and structure, investigating the ability of the proposed approaches for further applications would be interesting future work. For example, as the extracted information enables designers to have an accurate trace of the VPs' behavior, this ability can be used to assist the debugging process of SystemC-based VPs at the ESL. Thus, generating an automated debugging approach would also be a possible future work. Moreover, the extracted information provides designers with a detailed activity profile of the entire VP at different granularity levels (from IP cores down to single local variables), hence it could be used for power analysis of VPs at the ESL [68].

References

1. Accellera Systems Initiative. http://www.accellera.org/downloads/standards/systemc
2. R. Alur, D.L. Dill, A theory of timed automata. Theor. Comput. Sci. **126**, 183–235 (1994)
3. R. Alur, C. Courcoubetis, D. Dill, Model-checking in dense real-time. Inf. Comput. **104**(1), 2–34 (1993)
4. A. Ardeshiricham, W. Hu, J. Marxen, R. Kastner, Register transfer level information flow tracking for provably secure hardware design, in *Design, Automation and Test in Europe (DATE)* (IEEE, Piscataway, 2017), pp. 1691–1696
5. J. Aynsley (ed.), *OSCI TLM-2.0 Language Reference Manual*. Open SystemC Initiative (OSCI) (2009)
6. J. Aynsley, TLM-2.0 base protocol checker. https://www.doulos.com/knowhow/systemc/tlm2. Accessed 30 Jan 2018
7. M. Bawadekji, D. Große, R. Drechsler, TLM protocol compliance checking at the electronic system level, in *IEEE International Symposium on Design and Diagnostics of Electronic Circuits and Systems (DDECS)* (2011), pp. 435–440
8. D. Berner, J. Piere Talpin, H. Patel, D.A. Mathaikutty, E. Shukla, SystemCXML: an extensible SystemC front end using XML, in *Forum on Specification and Design Languages (FDL)* (2005), pp. 405–409
9. M.-M. Bidmeshki, Y. Makris, Toward automatic proof generation for information flow policies in third-party hardware IP, in *IEEE International Symposium on Hardware Oriented Security and Trust (HOST)* (IEEE, Piscataway, 2015), pp. 163–168
10. N. Blanc, D. Kroening, N. Sharygina, Scoot: a tool for the analysis of SystemC models, in *Tools and Algorithms for the Construction and Analysis of Systems (TACAS)* (Springer, Berlin, 2008), pp. 467–470
11. E. Bosman, A. Slowinska, H. Bos, Minemu: the world's fastest taint tracker, in *Recent Advances in Intrusion Detection (RAID)* (Springer, Berlin, 2011), pp. 1–20
12. H. Broeders, R. Van Leuken, Extracting behavior and dynamically generated hierarchy from SystemC models, in *Design Automation Conference (DAC)* (ACM, New York, 2011), pp. 357–362
13. Cadence Incisive (SimVision), information available on Cadence website. http///www.cadence.com. Accessed 30 Jan 2016
14. M. Carbin, M.C. Rinard, Automatically identifying critical input regions and code in applications, in *International Symposium on Software Testing and Analysis*, ISSTA (ACM, New York, 2010), pp. 37–48

© Springer Nature Switzerland AG 2020

M. Goli, R. Drechsler, *Automated Analysis of Virtual Prototypes at the Electronic System Level*, https://doi.org/10.1007/978-3-030-44282-8

15. M. Carbin, S. Misailovic, M.C. Rinard, Verifying quantitative reliability for programs that execute on unreliable hardware, in *ACM International Conference on Object Oriented Programming Systems Languages and Applications*, OOPSLA (ACM, New York, 2013), pp. 33–52

16. V.K. Chippa, D. Mohapatra, A. Raghunathan, K. Roy, S.T. Chakradhar, Scalable effort hardware design: exploiting algorithmic resilience for energy efficiency, in *Design Automation Conference (DAC)* (ACM, New York, 2010), pp. 555–560

17. V.K. Chippa, S.T. Chakradhar, K. Roy, A. Raghunathan, Analysis and characterization of inherent application resilience for approximate computing, in *Design Automation Conference (DAC)*, pp. 1–9 (2013)

18. C.N. Chou, Y.S. Ho, C. Hsieh, C.Y. Huang, Symbolic model checking on SystemC designs, in *Design Automation Conference (DAC)* (2012), pp. 327–333

19. J. Clause, W. Li, A. Orso, Dytan: a generic dynamic taint analysis framework, in *International Symposium in Software Testing and Analysis (ISSTA)* (ACM, New York, 2007), pp. 196–206

20. T. De Schutter, *Better Software. Faster!: Best Practices in Virtual Prototyping* (Synopsys Press, Mountain View, 2014)

21. S. Drzevitzky, M. Platzner, Achieving hardware security for reconfigurable systems on chip by a proof-carrying code approach, in *International Symposium on Reconfigurable Communication-centric Systems-on-Chip (ReCoSoC)* (IEEE, Piscataway, 2011), pp. 1–8

22. S. Drzevitzky, U. Kastens, M. Platzner, Proof-carrying hardware: towards runtime verification of reconfigurable modules, in *International Conference on Reconfigurable Computing and FPGAs (ReConFig)* (IEEE, Piscataway, 2009), pp. 189–194

23. Z. Du, K. Palem, A. Lingamneni, O. Temam, Y. Chen, C. Wu, Leveraging the error resilience of machine-learning applications for designing highly energy efficient accelerators, in *Asia and South Pacific Design Automation Conference (ASP-DAC)* (2014), pp. 201–206

24. W. Ecker, V. Esen, T. Steininger, M. Velten, M. Hull, Interactive presentation: implementation of a transaction level assertion framework in SystemC, in *Design, Automation and Test in Europe (DATE)* (2007), pp. 894–899

25. Y. Endoh, ASystemC: an AOP extension for hardware description language, in *International Conference on Aspect-oriented Software Development Companion, AOSD* (ACM, New York, 2011), pp. 19–28

26. H. Esmaeilzadeh, A. Sampson, L. Ceze, D. Burger, Neural acceleration for general-purpose approximate programs, in *Annual IEEE/ACM International Symposium on Microarchitecture* (2012), pp. 449–460

27. A. Ferraiuolo, R. Xu, D. Zhang, A.C. Myers, G.E. Suh, Verification of a practical hardware security architecture through static information flow analysis, in *SIGOPS Operating Systems Review* (2017), pp. 555–568

28. L. Ferro, L. Pierre, Isis: runtime verification of TLM platforms, in *Forum on Specification and Design Languages (FDL)* (2009), pp. 1–6

29. G. Fey, D. Große, T. Cassens, C. Genz, T. Warode, R. Drechsler, ParSyC: an efficient SystemC parser, in *Workshop on Synthesis and System Integration of Mixed Information technologies (SASIMI)* (2004), pp. 148–154

30. Freecores. https://github.com/freecores/. Accessed 30 Jan 2017

31. T.A.S. Foundation, Xerces C++ validating XML parse. http://xml.apache.org/xerces-c/. Accessed 30 Jan 2016

32. V. Ganesh, T. Leek, M. Rinard, Taint-based directed whitebox fuzzing, in *International Conference on Software Engineering (ICSE)* (IEEE Computer Society, Washington, 2009), pp. 474–484

33. C. Genz, R. Drechsler, System exploration of SystemC designs, in *IEEE Computer Society Annual Symposium on Emerging VLSI Technologies and Architectures (ISVLSI)* (2006), pp. 335–342

34. C. Genz, R. Drechsler, Overcoming limitations of the SystemC data introspection, in *Design, Automation and Test in Europe Conference Exhibition (DATE)* (2009), pp. 590–593

35. G. Gillani, A. Kokkeler, Improving error resilience analysis methodology of iterative workloads for approximate computing, in *Computing Frontiers Conference* (ACM, New York, 2017), pp. 374–379

36. M. Goli, R. Drechsler, Scalable simulation-based verification of SystemC-based virtual prototypes, in *Euromicro Conference on Digital System Design (DSD)* (2019), pp. 522–529

37. M. Goli, J. Stoppe, R. Drechsler, AIBA: an automated intra-cycle behavioral analysis for SystemC-based design exploration, in *IEEE International Conference on Computer Design (ICCD)* (2016), pp. 360–363

38. M. Goli, J. Stoppe, R. Drechsler, Automatic equivalence checking for SystemC-TLM 2.0 models against their formal specifications, in *Design, Automation and Test in Europe (DATE)* (2017), pp. 630–633

39. M. Goli, J. Stoppe, R. Drechsler, Automatic protocol compliance checking of SystemC TLM-2.0 simulation behavior using timed automata, in *IEEE International Conference on Computer Design (ICCD)* (2017), pp. 377–384

40. M. Goli, J. Stoppe, R. Drechsler, Automated non-intrusive analysis of electronic system level designs. IEEE Trans. Comput. Aided Des. Integr. Circuits Syst. **39**(2), 492–505 (2018). https://doi.org/10.1109/TCAD.2018.2889665

41. M. Goli, M. Hassan, D. Große, R. Drechsler, Automated analysis of virtual prototypes at electronic system level. in *Great Lakes Symposium on VLSI, (GLSVLSI)* (2019), pp. 307–310

42. M. Goli, J. Stoppe, R. Drechsler, Resilience evaluation for approximating SystemC designs using machine learning techniques, in *International Symposium on Rapid System Prototyping (RSP)* (2018), pp. 97–103

43. M. Goli, M. Hassan, D. Große, R. Drechsler, Security validation of VP-based SoCs using dynamic information flow tracking. Inf. Technol. **61**(1), 45–58 (2019). https://doi.org/10.1515/itit-2018-0027

44. D. Große, R. Drechsler, L. Linhard, G. Angst, Efficient automatic visualization of SystemC designs, in *Forum on Specification and Design Languages (FDL)* (2003), pp. 646–658

45. D. Große, H.M. Le, R. Drechsler: proving transaction and system-level properties of untimed SystemC TLM designs, in *ACM/IEEE International Conference on Formal Methods and Models for Codesign (MEMOCODE)* (2010), pp. 113–122

46. X. Guo, R.G. Dutta, Y. Jin, Eliminating the hardware-software boundary: a proof-carrying approach for trust evaluation on computer systems. IEEE Trans. Inf. Forensics Secur. **12**(2), 405–417 (2017)

47. V. Gupta, D. Mohapatra, S.P. Park, A. Raghunathan, K. Roy, Impact: imprecise adders for low-power approximate computing, in *IEEE/ACM International Symposium on Low Power Electronics and Design (ISLPED)* (2011), pp. 409–414

48. A. Habibi, S. Tahar, Design and verification of SystemC transaction-level models, in *IEEE Transactions on Very Large Scale Integration (VLSI) Systems*, vol. 14 (2006), pp. 57–68

49. M. Hassan, V. Herdt, H.M. Le, M. Chen, D. Große, R. Drechsler, Data flow testing for virtual prototypes, in *Design, Automation and Test in Europe (DATE)* (2017), pp. 380–385

50. M. Hassan, V. Herdt, H.M. Le, D. Große, R. Drechsler, Early SoC security validation by VP-based static information flow analysis, in *IEEE/ACM International Conference on Computer Aided Design (ICCAD)* (2017), pp. 400–407

51. P. Herber, S. Glesner, A HW/SW co-verification framework for SystemC. ACM Trans. Embed. Comput. Syst. **12**, 61:1–61:23 (2013)

52. P. Herber, J. Fellmuth, S. Glesner, Model checking SystemC designs using timed automata, in *IEEE/ACM/IFIP International Conference on Hardware/Software Codesign and System Synthesis, CODES+ISSS* (ACM, New York, 2008), pp. 131–136

53. V. Hodge, J. Austin, A survey of outlier detection methodologies, in *Artificial Intelligence Review (AIR)* (2004), pp. 85–126

54. http://www.inf.ed.ac.uk/teaching/courses/inf2c-cs/13-14/labs/lab2.html. Accessed 30 Jan 2017

55. W. Hu, J. Oberg, A. Irturk, M. Tiwari, T. Sherwood, D. Mu, R. Kastner, Theoretical fundamentals of gate level information flow tracking. IEEE Trans. Comput. Aided Des. Integr. Circuits Syst. **30**(8), 1128–1140 (2011)

56. IEEE Standard SystemC Language Reference Manual, IEEE Std 1666–2005 (2006), pp. 1–423

57. Y. Jin, B. Yang, Y. Makris, Cycle-accurate information assurance by proof-carrying based signal sensitivity tracing, in *IEEE International Symposium on Hardware Oriented Security and Trust (HOST)* (IEEE, Piscataway, 2013), pp. 99–106

58. M. Kallel, Y. Lahbib, R. Tourki, A. Baganne, Verification of SystemC transaction level models using an aspect-oriented and generic approach, in *International Conference on Design Technology of Integrated Systems in Nanoscale Era (DTIS)* (2010), pp. 1–6

59. F. Karlsruhe, Kascpar - karlsruhe SystemC parser suite (2012)

60. A. Kaushik, H.D. Patel, SystemC-clang: an open-source framework for analyzing mixed-abstraction SystemC models, in *Forum on Specification and Design Languages (FDL)* (2013), pp. 1–8

61. V.P. Kemerlis, G. Portokalidis, K. Jee, A.D. Keromytis, libdft: Practical dynamic data flow tracking for commodity systems. ACM SIGPLAN Not. **47**(7), 121–132 (2012)

62. H. Khattri, N.K.V. Mangipudi, S. Mandujano, HSDL: A security development lifecycle for hardware technologies, in: *IEEE International Symposium on Hardware Oriented Security and Trust (HOST)* (IEEE, Piscataway, 2012), pp. 116–121

63. W. Klingauf, M. Geffken, Design structure analysis and transaction recording in SystemC designs: a minimal-intrusive approach, in *Forum on Specification and Design Languages (FDL)* (2006)

64. P. Kocher, D. Genkin, D. Gruss, W. Haas, M. Hamburg, M. Lipp, S. Mangard, T. Prescher, M. Schwarz, Y. Yarom, Spectre attacks: exploiting speculative execution. Preprint. arXiv:1801.01203 (2018)

65. C. Lattner, LLVM and Clang: next generation compiler technology, in *BSDCan 2008: The BSD Conference* (2008), pp. 1–2

66. H.M. Le, D. Große, V. Herdt, R. Drechsler, Verifying SystemC using an intermediate verification language and symbolic simulation, in *Design Automation Conference (DAC)* (2013), pp. 1–6

67. J. Lee, A. Shrivastava, Static analysis to mitigate soft errors in register files, in *Design, Automation and Test in Europe (DATE)* (2009), pp. 1367–1372

68. D. Lemma, M. Goli, D. Große, R. Drechsler, Power intent from initial ESL prototypes extracting power management parameters, in *IEEE Nordic Circuits and Systems Conference (NORCAS)* (2018), pp. 1–6

69. X. Li, M. Tiwari, J.K. Oberg, V. Kashyap, F.T. Chong, T. Sherwood, B. Hardekopf, Caisson: a hardware description language for secure information flow. ACM SIGPLAN Conf. Program. Lang. Design Implement. **46**(6), 109–120 (2011)

70. X. Li, V. Kashyap, J.K. Oberg, M. Tiwari, V.R. Rajarathinam, R. Kastner, T. Sherwood, B. Hardekopf, F.T. Chong, Sapper: a language for hardware-level security policy enforcement. ACM SIGARCH Comput. Archit. News **42**(1), 97–112 (2014)

71. M. Lipp, M. Schwarz, D. Gruss, T. Prescher, W. Haas, S. Mangard, P. Kocher, D. Genkin, Y. Yarom, M. Hamburg, Meltdown. CoRR, abs/1801.01207 (2018)

72. S. Liu, K. Pattabiraman, T. Moscibroda, B.G. Zorn, Flikker: saving dram refresh-power through critical data partitioning. SIGPLAN Not. **47**, 213–224 (2011)

73. E. Love, Y. Jin, Y. Makris, Proof-carrying hardware intellectual property: a pathway to trusted module acquisition. IEEE Trans. Inf. Forensics Secur. **7**(1), 25–40 (2012)

74. K. Marquet, M. Moy, Pinavm: a SystemC front-end based on an executable intermediate representation, in *Embedded software (EMSOFT)* (ACM, New York, 2010), pp. 79–88

75. G. Martin, B. Bailey, A. Piziali, *ESL Design and Verification: A Prescription for Electronic System Level Methodology* (Morgan Kaufmann Publishers Inc., San Francisco, 2007)

76. S.A. Metwalli, Y. Hara-Azumi, SSA-AC: static significance analysis for approximate computing. ACM Trans. Des. Autom. Electron. Syst. **24**(3), 34:1–34:17 (2019)

77. S. Mittal, A survey of techniques for approximate computing, in *ACM Computing Surveys (CSUR)* (2016), p. 62
78. D. Mohapatra, V.K. Chippa, A. Raghunathan, K. Roy, Design of voltage-scalable meta-functions for approximate computing, in *Design, Automation and Test in Europe (DATE)* (2011), pp. 1–6
79. G.E. Moore, Progress in digital integrated electronics, in *International Electron Devices Meeting*, vol. 21 (1975), pp. 11–13
80. M. Moy, F. Maraninchi, L. Maillet-Contoz, Pinapa: an extraction tool for SystemC descriptions of systems-on-a-chip, in *Embedded Software (EMSOFT)* (ACM, New York, 2005), pp. 317–324
81. Z. Navabi, *VHDL: Analysis and Modeling of Digital Systems*, 1st ed. (McGraw-Hill, Inc., New York, 1992)
82. Z. Navabi, *VERILOG Digital System Design: Analysis and Design of Digital Systems with Cdrom*. (McGraw-Hill, Inc., New York, 1999)
83. K. Nepal, Y. Li, R.I. Bahar, S. Reda, Abacus: a technique for automated behavioral synthesis of approximate computing circuits, in *Design, Automation Test In: Europe Conference Exhibition (DATE)* (2014), pp. 1–6
84. B. Nongpoh, R. Ray, S. Dutta, A. Banerjee: Autosense: a framework for automated sensitivity analysis of program data. IEEE Trans. Softw. Eng. **43**(12), 1110–1124 (2017)
85. Novas Verdi, information available on Novas website. http///www.novas.com. Accessed 30 Jan 2016
86. orahyn: Lzw-encoder. https://github.com/arshadri/lzw_systemc/tree/master/systemc. Accessed 30 Jan 2017
87. T. Parr, *Language Translation Using PCCTS and c++: A Reference Guide* (Automata Publishing Co., San Jose, 1997)
88. H.D. Patel, D.A. Mathaikutty, D. Berner, S.K. Shukla, Systemcxml: an extensible SystemC front end using XML. Technical Report (2005)
89. L. Pierre, M. Chabot, Assertion-based verification for SoC models and identification of key events, in *EUROMICRO Digital System Design Conference (DSD)* (2017), pp. 54–61
90. F. Qin, C. Wang, Z. Li, H.-s. Kim, Y. Zhou, Y. Wu, Lift: a low-overhead practical information flow tracking system for detecting security attacks, in *Annual IEEE/ACM International Symposium on Microarchitecture (MICRO)* (IEEE, Piscataway, 2006), pp. 135–148
91. M. Rinard, Probabilistic accuracy bounds for fault-tolerant computations that discard tasks, in *Annual International Conference on Supercomputing, ICS* (ACM, New York, 2006), pp. 324–334
92. M.C. Rinard, Using early phase termination to eliminate load imbalances at barrier synchronization points. SIGPLAN Not. **42**, 369–386 (2007)
93. P. Roy, R. Ray, C. Wang, W.F. Wong, ASAC: automatic sensitivity analysis for approximate computing, in *SIGPLAN/SIGBED Conference on Languages, Compilers and Tools for Embedded Systems*, LCTES (ACM, New York, 2014), pp. 95–104
94. J. Rumbaugh, I. Jacobson, G. Booch (eds.), *The Unified Modeling Language Reference Manual* (Addison-Wesley Longman Ltd., Boston, 1999)
95. A. Sampson, W. Dietl, E. Fortuna, D. Gnanapragasam, L. Ceze, D. Grossman, EnerJ: approximate data types for safe and general low-power computation, in *ACM SIGPLAN Conference on Programming Language Design and Implementation, PLDI* (ACM, New York, 2011), pp. 164–174
96. A. Sampson, J. Nelson, K. Strauss, L. Ceze, Approximate storage in solid-state memories, in *Annual IEEE/ACM International Symposium on Microarchitecture*, MICRO (ACM, New York, 2013), pp. 25–36
97. B.C. Schafer, A. Mahapatra: S2CBench: Synthesizable SystemC benchmark suite for high-level, in *IEEE Embedded Systems Letters*, vol. 3 (2014), pp. 53–56
98. T. Schmidt, G. Liu, R. Dömer, Automatic generation of thread communication graphs from SystemC source code, in *Software and Compilers for Embedded Systems (SCOPES)* (2016), pp. 108–115

99. T. Schubert, W. Nebel, The quiny SystemCTM front end: Self-synthesising designs, in *Selected Contributions from Forum on specification and Design Languages (FDL)* (Springer, Berlin, 2007), pp. 93–109

100. C. Schulz-Key, M. Winterholer, T. Schweizer, T. Kuhn, W. Rosentiel, Object-oriented modeling and synthesis of SystemC specifications, in *Asia and South Pacific Design Automation Conference (ASP-DAC)* (2004), pp. 238–243

101. T. Schuster, R. Meyer, R. Buchty, L. Fossati, M. Berekovic, Socrocket – A virtual platform for the European space agency's SOC development, in *Reconfigurable Communication-Centric Systems-on-Chip (ReCoSoC)* (2014), pp. 1–7. Available at http://github.com/socrocket

102. S. Skorobogatov, C. Woods, Breakthrough silicon scanning discovers backdoor in military chip, in *International Conference on Cryptographic Hardware and Embedded Systems (CHES)* (Springer, Berlin, 2012), pp. 23–40

103. W. Snyder: SystemPerl homepage.http://www.veripool.com/systemperl.html. Accessed 30 Jan 2016

104. H. Sohofi, Z. Navabi, Assertion-based verification for system-level designs, in *International Symposium on Quality Electronic Design (ISQED)* (2014), pp. 582–588

105. R. Stallman, C. Support, *Debugging with GDB: The GNU Source-Level Debugger*, 9th edn. (Free Software Foundation, Boston, 2010)

106. J. Stoppe, R. Drechsler, Analyzing SystemC designs: SystemC analysis approaches for varying applications. Sensors **15**(5), 10399–10421 (2015)

107. J. Stoppe, R. Wille, R. Drechsler, Data extraction from SystemC designs using debug symbols and the SystemC API, in *IEEE Computer Society Annual Symposium on Emerging VLSI Technologies and Architectures (ISVLSI)* (IEEE, Piscataway, 2013), pp. 26–31

108. G.E. Suh, J.W. Lee, D. Zhang, S. Devadas, Secure program execution via dynamic information flow tracking. ACM Sigplan Not. **39**(11), 85–96 (2004)

109. D. Tabakov, M.Y. Vardi, Monitoring temporal SystemC properties, in *ACM/IEEE International Conference on Formal Methods and Models for Codesign (MEMOCODE)* (2010), pp. 123–132

110. D. Tabakov, M.Y. Vardi, Automatic aspectization of SystemC, in *Workshop on Modularity in Systems Software, MISS* (ACM, New York, 2012), pp. 9–14

111. T.C. Team: Clang: a C language family frontend for LLVM. https://clang.llvm.org/. Accessed 10 Jan 2017

112. TechPowerUp, "Intel xeon platinum 8180." https://www.techpowerup.com/cpudb/2055/xeon-platinum-8180. Accessed 12 Jan 2018

113. O. Temam, A defect-tolerant accelerator for emerging high-performance applications, in *International Symposium on Computer Architecture (ISCA)* (IEEE Computer Society, Washington, 2012), pp. 356–367

114. M. Tiwari, H.M. Wassel, B. Mazloom, S. Mysore, F.T. Chong, T. Sherwood, Complete information flow tracking from the gates up. ACM Sigplan Not. **44**(3), 109–120 (2009)

115. N. Vachharajani, M.J. Bridges, J. Chang, R. Rangan, G. Ottoni, J.A. Blome, G.A. Reis, M. Vachharajani, D.I. August, Rifle: an architectural framework for user-centric information-flow security, in *Annual IEEE/ACM International Symposium on Microarchitecture (MICRO)* (IEEE, Piscataway, 2004), pp. 243–254

116. F. Vahid, *Digital Design with RTL Design, Verilog and VHDL*, 2nd ed. (Wiley Publishing, Hoboken, 2010)

117. D. van Heesch, Doxygen: source code documentation generator tool. http://www.doxygen.org. Accessed 30 Jan 2016

118. V. Vassiliadis, J. Riehme, J. Deussen, K. Parasyris, C.D. Antonopoulos, N. Bellas, S. Lalis, U. Naumann, Towards automatic significance analysis for approximate computing, in *International Symposium on Code Generation and Optimization*, CGO (ACM, New York, 2016), pp. 182–193

119. W. Xu, S. Bhatkar, R. Sekar, Taint-enhanced policy enforcement: a practical approach to defeat a wide range of attacks, in *USENIX Security Symposium* (2006), pp. 121–136

120. D. Zhang, Y. Wang, G.E. Suh, A.C. Myers, A hardware design language for timing-sensitive information-flow security. ACM SIGPLAN Not. **50**(4), 503–516 (2015)

Index

© Springer Nature Switzerland AG 2020
M. Goli, R. Drechsler, *Automated Analysis of Virtual Prototypes at the Electronic
System Level*, https://doi.org/10.1007/978-3-030-44282-8

Printed in the United States
by Baker & Taylor Publisher Services